PRAISE FOR *Fringe-ology*

"*Fringe-ology* is one of the most refreshing books in years about so-called paranormal phenomena. Volk transcends the hardened, militant bombast of skeptics who refuse to give an inch, as well as the accepting attitudes of true believers whose critical faculties have deserted them. Volk is open to the great mysteries and strangeness of human experience, and he is willing to let the chips fall where they may. And did I mention this book is pure fun?"

—Larry Dossey, M.D., author of *The Power of Premonitions* and *Reinventing Medicine*

"Skeptical investigative journalist Steve Volk explores the paranormal and finds—to his astonishment—the leading edge of science. *Fringe-ology* is a thoughtful, myth-shattering exploration of discoveries that may be in the process of transforming the paranormal into normal."

—Dean Radin, Ph.D., senior scientist, Institute of Noetic Sciences

"Steve Volk has written a fascinating overview of several taboos that defy explanations by mainstream science. He illuminates these fringes of human behavior through scientific data as well as case studies. *Fringe-ology* shows that he has clearly done his homework, and where the hard data contradict the beliefs of "true believers," he lets the reader know. Unlike many other books on the topic, *Fringe-ology* has a well-developed storyline that keeps up the suspense. From near-death experiences to after-death communications, from haunting to hypnosis, from creative dreams to quantum physics, this book will not make all of its readers happy but it does contain something of interest to everyone."

—Stanley Krippner, Ph.D., professor of psychology, Saybrook University, coeditor of *Debating Psychic Experience*

"*Fringe-ology* brings a poet's eye to the frayed edges between the known and unknown, belief and skepticism. A beautifully written, gripping read filled with humor, wonder, and vulnerability, *Fringe-ology* is a dive into the paranormal that even a hardcore skeptic like myself can enjoy."

—Mat Johnson, author of *Pym*

"Volk takes readers on a roller-coaster ride through what is arguably the most transforming and thrilling period in world history—when science and mysticism merge to become the new reality. What more could you ask for from a book? Volk is smart, persuasive, humorous, and always entertaining. Just be sure to strap on your seatbelt."

—Sidney Kirkpatrick, author of *Edgar Cayce: An American Prophet*

"*Fringe-ology* is just an astonishingly good book on the subject of the supernatural—smart, funny, insightful, and wise. Partly a personal quest, partly a journey through our often inexplicable and occasionally eerie universe, Volk's tale will leave you wondering about what's real and what isn't and, ultimately, how we define reality at all."

—Deborah Blum, author of *Ghost Hunters: William James and the Search for Scientific Proof of Life After Death*

"The paranormal community owes Steve Volk a debt of gratitude. He has ventured where few mainstream investigative journalists have dared to go before. By his immersion research into various anomalies, from ESP to deathbed visions, with the open-minded but critical eye of a reporter, he has uncovered surprising evidence for much that the average reader may have considered impossible. Presented in a way that is both engaging and entertaining, his chapters will resonate with the reader long after the book is finished, perhaps even provoking a lucid dream. This is a must read for all those seeking life's 'possibilities.'"

—Sally Rhine Feather, executive director emeritus, Rhine Research Center, coauthor of *The Gift*.

"*Fringe-ology* is great, and as an introduction to science/religion topics it is one of the best I've read."

—Robert McLuhan, author of *Randi's Prize*.

"Has science really shut the door on ghosts, telepathy, and other paranormal phenomena? Not at all, as Steve Volk shows in this mind-bending work of journalism. *Fringe-ology* takes us for a walk on the wild side and explains why—when we're dealing with experiences that are truly unexplainable—the most honest response is often simply to say 'I don't know.'"

—Steve Paulson, executive producer, *To the Best of Our Knowledge*

FRINGE-OLOGY

FRINGE-OLOGY

HOW I TRIED TO EXPLAIN AWAY THE UNEXPLAINABLE—AND COULDN'T

Steve Volk

HarperOne
An Imprint of HarperCollins*Publishers*

HarperCollins books may be purchased for educational, business, or sales promotional use. For information, please e-mail the Special Markets Department at SPsales@harpercollins.com.

HarperCollins website: http://www.harpercollins.com

HarperCollins®, ®, and HarperOne™ are trademarks of HarperCollins Publishers

FIRST HARPERCOLLINS PAPERBACK EDITION PUBLISHED IN 2012

Designed by Laura Lind Design

Library of Congress Cataloging-in-Publication Data
Volk, Steve.
Fringe-ology : how I tried to explain away the unexplainable—and couldn't / Steve Volk. — 1st ed.
p. cm.
ISBN 978–0–06–185772–0
1. Parapsychology. 2. Supernatural. I. Title.
BF1031.V65 2011
130—dc22 2010048655

HB 06.07.2018

To my parents, Gerald and Joanne Volk,
for absolutely everything.

CONTENTS

FRINGE-OLOGY

INTRODUCTION

What We Talk About When We Talk About the Paranormal

Or: Why we can't even agree on just what it is we're discussing

My point is not that religion itself is the motivation for wars, murders and terrorist attacks, but that religion is the principal label, and the most dangerous one, by which a "they" as opposed to a "we" can be identified at all.

—Richard Dawkins, "Time to Stand Up"

It's cramped and irrational to say that there is no God—and premature. Because we are pathetically ignorant of the universe.

—Martin Amis

People, I just want to say, you know, can we all get along?

—Rodney King

We have been sitting in the dark for hours. I am on the outside of a circle of a half-dozen strangers, reclining on the thick carpeting of a suburban home, watching the silhouette of a hulking man with a digital audio recorder. We are waiting, and listening, for some proof of a ghostly presence in the room. And I, for one, am not optimistic. I've come here with Lou Gentile, a well-known figure within the murky realms of ghost hunting.

And so far, we have found nothing to confirm what this family in central New Jersey has told us: they tell us about strange rappings, jiggling doorknobs, and an occasional bang in the basement. We have heard nothing of the kind. But more than that, we both think the vibe this family gives off is strange enough without the additive of spirits.

The family patriarch, who I'll refer to here as "Paul," moves with the disassociated air of a ghost himself. He is a quiet, intense divorcé who seems to rule over his girlfriend and son with his ominous silences. Like me, he has recently lost a parent—in this case, his father. And while his family seems to maintain an appropriate skepticism, he clearly wants to believe.

Earlier, he showed Gentile a series of odd photos, including one that displayed what appeared to be a large, jagged light in his bedroom. He seemed happy when Gentile told him he wasn't sure what kind of camera defect might produce that anomaly. "It could be a defect," Gentile told him, "but I haven't seen one quite like this."

Paul also showed us a video he made in the basement, which was about as interesting, aesthetically, as might be expected. The static image he captured was mostly darkness, with the outline of a weight bench in the middle ground and still more darkness fanning out behind it. Every few seconds, however, a ping echoed in the room. "The sound is probably distorted by the condenser microphone on your camera," Gentile told him. "But it's worth investigating."

Gentile sent me down to the basement at that point, to sit in the dark. And though I was suitably scared for a minute or so, it quickly became clear to me that the "mysterious noises" Paul drew our attention to were produced by nothing more spectral than the air conditioning ducts that cut back and forth across the ceiling. I reported my findings to Gentile, who expected as much. "It doesn't mean nothing is happening here," he told Paul. "But the sounds on the video are just produced by your air conditioning unit."

Paul, sitting cross-legged on the floor, looked distraught. Gentile readied his little digital recorder and asked Paul to turn off all the lights in the house. Hours passed. Gentile asked questions into the darkness, then played back the audio, listening for ghostly responses. Believers call this "electronic voice phenomenon." I considered it a kind of investigative dead end, a series of unintelligible, scratchy noises that could easily have been produced by the recorder itself—the sounds occasionally coalescing by chance into a snatch that could be mistaken for a word or maybe even a phrase. On this night, Gentile didn't seem particularly impressed by the results either. So to shake things up, he invited everyone in the circle to take turns asking questions. When it was Paul's turn, he knelt down in front of the recorder and asked the only question of the dark that could be expected from a grieving son: "Are you my father?"

Gentile's recorder captured no response.

|||||||||||

FOR MANY PEOPLE, AND especially hard-core skeptics, all the paranormal ever amounts to is the wishful thinking of the grieving. By this time, however, I already knew better. I had been out with Gentile on more than a half-dozen investigations, and while much of what we saw and heard could be easily tracked to some earthly origin, some couldn't. Further, while some of the self-proclaimed witnesses seemed to like the idea of ghosts in their houses, others were hoping Gentile might find some prosaic explanation. This in and of itself should not come as much of a surprise. If all our strange experiences could be explained so easily as a simple mishmash of wishful thinking and creaking floorboards, we'd have no need of the word *paranormal*. But we do. Paranormal experiences have been with us since the beginning of recorded history. And they don't seem inclined to go away any time soon.

In fact, just on the subject of spirits, researcher and folklorist Lionel Fanthorpe announced in 2010 that ghost sightings were at their highest point in twenty-five years. We continue to be inundated with tales of unidentified flying objects (UFOs), and some of the accounts could fuel a feature film: Winston Churchill supposedly quashed a dramatic UFO report, fearing a public panic; famous U.K. computer hacker Gary McKinnon went poking around in NASA's mainframe for evidence of alien spacecraft, evidence that he claims he found; but alas, he had no time to download the telltale photograph before his connection was cut off by an angry American government.

But the question we need to ask going forward is, What exactly are we talking about when we talk about the paranormal? I could, in fact, fill endless pages with odd, tantalizing stories. Did you hear the one about the cab drivers on the Solomon Islands? In the fall of 2008 they started complaining about picking up passengers who acted completely normal until they disappeared suddenly from the backseat. And they never did pay their fares. In the coming pages, I write about near-death experiences (NDEs), mental telepathy, quantum consciousness, UFOs, a mystic astronaut, ghost hunting, and a pair of scientists doing their level best to study aspects of human experience often derided as paranormal. But this book is about more than any of these things. This is a book about us.

We live in a world of false certainties: Whether we are discussing politics, religion, or economics, when we flip on our televisions or open our Web browsers to a news site, we encounter the often strongly held opinions of others—opinions that lead us into a series of binary choices: conservative or liberal, believer or atheist, capitalist or socialist. My argument, simply, is that these are false choices—that there are middle paths that bear more fruit. But unfortunately, as we'll see in the pages of this book, it is human nature and an automatic function of the brain to frame conflicting worldviews in extreme terms of right and wrong, good and evil, rational and irrational. And I would argue that this kind of Us Versus Them thinking is perhaps best and most readily seen in debates about the paranormal.

The word *paranormal* is itself a kind of victim of human psychology, too often conflated with *supernatural*: "of or relating to an order of existence beyond the visible observable universe"; especially*: "of or relating to God or a god, demigod, spirit, or devil." Paranormal*, conversely, can be and often is defined in far broader and more scientifically useful terms: "of or pertaining to events or perceptions occurring without scientific explanation." In fact, *Webster's Third New International Dictionary* defines *paranormal* as "beyond the range of scientifically known or recognizable phenomena." If we take these definitions, the supernatural seems to force us toward religion, while the paranormal merely forces us to say, "I don't know." There should be no shame in that, but I think the faithful too often want to equate their beliefs with knowledge, while the skeptics fear that admitting a lack of a final answer opens the door to all manner of hoo ha, including God. The skeptics also tend to view the words *supernatural* and *paranormal* as if they are easily interchangeable, but whereas the supernatural seems to lie firmly beyond science, the paranormal waits patiently for the technology and the willing scientists necessary for its discovery.

Whether we consider God in this equation or not, such a definition of paranormal as simply representing the unknown is striking because it encompasses a whole variety of phenomena we don't normally consider paranormal, like the placebo effect. It is beyond strange that people experience healing effects after being fed a useless sugar pill, but it happens. Researchers have pinned the cause to belief, but the exact mechanism of how belief in the efficacy of a medicine not actually administered leads to a reduction of asthma symptoms, for instance, remains unknown. Pharmaceutical companies are hoping to figure it all out, because incredibly, the placebo effect is getting stronger and no one is sure why. Is the placebo effect paranormal? Or belief? Hypnotism is another example of a phenomenon we don't normally talk about when we talk about the paranormal. And unlike the placebo effect, some skeptics try to deny hypnosis even exists, attributing its effects to the "power of suggestion." There is reason for us to want to bury hypnotists in a great big hole: if careful procedures aren't followed, recovered memories are often false, and people have been wrongfully prosecuted based on "evidence" gleaned from hypnosis. But there are numerous, well-documented cases of hypnotism causing incredible relief from pain. Hypnotism has been dramatically effective in controlling the pain of surgery and childbirth. So if hypnotism isn't real, something dramatic is still occurring under that name. And scientists, again, have thus far failed to explain what it is or precisely how it works. Is hypnotism paranormal?

It is time for us to broaden our minds, lift the stigma from the word *paranormal,* and instead see the opportunities these odd stories present for us: in terms of science, the pursuit of the paranormal has propelled civilization toward some of its most important advancements. The study of alchemy—the power to transmute lead into gold—gave rise to modern chemistry.

Francis Aston used the predictions of two occultists, which proved a springboard to discovering the isotope—a link long omitted from scientific textbooks. And then there is Hans Berger, who sought a physical mechanism for psychic events and landed on electricity as the answer. Berger invented the electroencephalogram, or EEG, to measure theretofore unheard-of electrical activity in the brain—and provided the foundation of modern neuroscience. In short, what is today seen as wacky often leads to tomorrow's progress. But in seeing the paranormal as grounds for investigation rather than argument, there is another, more immediate benefit we might get: to reconnect, each to the other, in a shared understanding that we are enjoined in a common predicament. At the most fundamental level of reality, we are fellow travelers stuck to the hull of a rock floating through space, without final answers to the questions that are traditionally most important to us: *What happens when we die? Is there a god? Are we alone in the universe? Why are we here?*

All we really know, in an epistemological sense, is that we *are* going to die. How nice. This is the existential reality of the human condition. But the paranormal potentially offers us answers. Because if ghost stories are merely folk tales, the product of overactive imaginations trying to deal with the existential angst of human experience, as skeptics maintain, they are worth writing down as documents of our innermost selves. And if some aspects of the paranormal prove to be real, as mystics contend, then we will have made a discovery worth inventing new words for—a discovery of *ginormous* proportions. Toward the end of this book, in fact, we'll find that such a discovery—a couple such discoveries—have perhaps already been made. As a result, I argue, taking the paranormal seriously means we gain a greater understanding of the world, regardless of the outcome. People have reported anomalous experiences since literally the beginning of recorded history. Plato told the story of Er, a soldier who awoke upon his own unlit funeral pyre, descended from that stack of sticks and spoke of a trip into the afterlife. President Abraham Lincoln famously dreamt of his own assassination. Coincidence, or did the president really see it coming? Psychiatrists Colin Ross and Shaun Joshi claim paranormal experiences are so common throughout the population that psychology must account for them in order to be comprehensive.

It is time we take a firm accounting of these stories. And I have my own reasons for conducting this research. I am, of course, interested in the fundamental questions of our existence, expressed in the deep panic of an existential crisis: *whyarewehere? arewealone? whathappenswhenwedie?* But the source of my interest is also more particularized.

I have made my living as a reporter for a dozen years now. And standard operating procedure for any journalist is to play the paranormal for laughs. Reporters assigned to Halloween "ghost hunt" stories go out with a local crew of amateur ghost seekers. They spend a couple of hours in a supposedly haunted location. Nine times out of ten nothing remotely unexplainable happens, and the reporter files a story that pokes gentle fun at the "real-life ghostbusters" who interpret every dip in temperature as a disembodied spirit. But I can't write that story, or at least I can't write that story in good conscience.

My own family passed a series of ghost stories on to me, stories they swore to be true. And I have my own memories of the events they describe. I remember the banging noises, for instance, that sounded like something was trying to hack its way in through the roof. I remember my older brother throwing his hands on top of his head and staring up at the ceiling in dismay. I remember my sisters claiming that their bed covers had been jerked from them, violently, as if by invisible hands.

My mother and father once sat and told me about what they called "the ghost." I was in my mid-twenties, recently graduated from college. I had instigated the conversation by telling them I wanted the definitive account. The tale they told me was sensational, complete with a spellbinding ending that seemed traumatic to them all those years later. While I'll save the details until chapter 5, my parents' reaction to the telling of them gripped me. They held hands across the kitchen table. They looked sick to their stomachs. They were more than twenty years removed from the events they recounted, yet still frightened.

Of course, this should not be. As a child of the Enlightenment, I should believe in a rational, materialistic explanation for everything that happens. So should they. By this logic, my parents would either be liars or fools for believing there might have been a ghost in our house. That assessment may sound harsh. But Western science doesn't yield easily to the paranormal. We live in an age when the turf wars between mystics and rationalists are growing considerably hotter. These days, in fact, atheists advocate for their point of view with the passion of priests looking for penitents. And what gets lost in all this controversy is a whole world of phenomena ripe for psychological and scientific exploration. What also gets lost is the essential human drama of it all, the husband and wife sitting at the table, white faced, after twenty years in the wake of an event they cannot explain.

This book seeks to demonstrate that the paranormal can be covered in the same way as any other story—and that, in fact, these subjects are among the most important we *could* cover, yielding fascinating narratives that speak to the most fundamental concerns humanity has faced. So, where possible, I assess the veracity of the strange tales people tell. And what I'll find is that some paranormal claims are far more possible than the skeptics like to admit. But my focus remains on the experience most accessible to journalism: the experience of confronting a mystery.

People who undergo paranormal experiences risk falling into a special, unofficial subclass of U.S. citizens—a subclass I call "Embarrassed Americans." For my chapter on UFO sightings (chapter 4), in fact, I make a trip to a town filled with the chagrined: Stephenville, Texas—dairy cow country. In January 2008, residents all over the county saw strange lights in the sky. But fewer than a dozen ever went public with their stories. The opportunity was there: swarms of national media descended upon the town. But many residents claimed they were embarrassed for themselves and their community. And all they had done for their trouble was see something in the sky that they couldn't explain.

The central problem, I think, is that as a species we *seem* to lack humility. One of the strangest books of recent times, *The End of Science*, was penned by John Horgan—who in spite of so much evidence to the contrary argues that science will not produce any more truly revolutionary findings. Of course, potential revolutions still abound: dark matter and dark energy are thought to make up roughly 95 percent of the universe, yet direct evidence of their existence eludes us. (Does that make 95 percent of the universe paranormal?)

Temple University recently produced a book that identifies thirteen competing theories of quantum mechanics and explores their implications for how we understand the very nature of existence. Some highly esteemed scientists, including Karl Pribram and quantum mechanics mastermind David Bohm, advanced the idea that the human mind and the entire universe are like a hologram—a three-dimensional projection from some other, more fundamental reality.

If the universe doesn't seem quite weird enough for you yet, consider the matter of time, a particularly sticky wicket: Einstein himself called time a "stubbornly persistent illusion" because, from the perspective of a physicist, there seems to be no obvious explanation for why we experience time in the linear fashion that we do. To explore the subject, physicists Yakir Aharonov and Jeff Tollaksen devised an incredible experiment, in which the act of measuring a particle at, say, 3:00 P.M., predictably changes the value of the same particle at, um, 2:30 P.M.—a half-hour earlier. Numerous labs around the world have been successfully conducting and replicating the experiment, which seems to indicate something awfully wild about reality: an action taken in the future can affect what happens in the present, at least at a subatomic level.

Aharonov and Tollaksen aren't sure precisely what to make of their own findings. But this is precisely the spot at which we can use a real, scientific mystery to understand something about ourselves and how we react to the paranormal. Most likely, you rebelled, internally, during this last paragraph. You balked at the idea that what we do at 3:00 P.M. can effect what happened at 2:30 P.M. But without belaboring the nature of time, there is a part of your brain that probably sent you a tremulous message to watch out when I wrote something that seems so nonsensical. Maybe you furrowed your eyebrows, your pulse quickened, you momentarily held your breath or even felt angry or dismissive, as if what I had written must be false and I must be stupid or even craven to write it. But here's the thing: that wasn't you, or at least not the rational, reasonable you. That was your brain talking—most dramatically, your amygdala, a necessary but frustrating part of the brain that we'll meet throughout this book. The amygdala is the spot in the brain I accuse of making us seem to lack humility—the part of our brain that can cause us to haughtily dismiss information we find threatening or don't understand.

When our place on the food chain was not so secure, and we had to deal with predator cats on a regular basis, the amygdala—a pair of almond-shaped structures near the base of our temporal lobes—did great work. Our brain processed visual images of a shadow moving in the grass, and our amygdala shouted, "Danger!" In response, we froze. Our more logical information-processing centers kicked in, quickly trying to determine: Is this shadow a crouching tiger or a hidden rabbit? If the shadow was big enough, our logical frontal lobes responded, *Close enough to a tiger for me*, our amygdala sent a stronger signal of abject fear in return, and we ran.

Millions of years later, *Homo sapiens* is here—and we brought our amygdalas with us. Some of us, like kids in the inner city, or soldiers in the battlefield, still need them a lot. These are people who worry on a daily basis about potent threats to their health—about a lump in a stranger's pocket that might mean he is carrying a handgun; about a mound of dirt on the side of the road which might cover a bomb. But for most of us, the amygdala (along with other parts of the brain responsible for mediating emotion and processing conflicting information) is responding to far less grave mysteries and instead sending us messages of anxiety and fear whenever necessary and much of the time besides, including when the boss says something harsh to us at work, a coworker cuts us a nasty look, or when we hear an idea that conflicts with our worldview.

This has profound implications for all of us, and our conversations about the paranormal, and means oftentimes our first reaction, even if it is about an intellectual subject, is an emotional one: We react to the ideas we hear not only with our rational frontal lobes but with this primitive part of our brain. And when we feel emotionally committed to a position, that is precisely the time we're in the greatest danger of reacting—not from our frontal lobes, like enlightened human beings, but from our amygdalas, like angry or frightened monkeys. Believers sometimes consider those of no or different faith downright unholy. Nonbelievers, of late, take great delight in openly deriding believers as irrational and childlike. And too often the rest of us wind up listening to people letting their amygdalas inspire far too much of the talking.

There is intriguing research that backs this up, including not just brain scan technology but terror management theory. It seems that both religious and irreligious people see death and threats to their worldviews as of a piece; in other words, he who threatens my life and he who threatens my way of looking at the world are, on a psychological level, related. That is a dangerous way of thinking, but it seems we're built for it—machines constructed to fight.

As we'll see late in this book, mystics have come up with some great ways of taming the more unruly, emotional centers of our brain, including the amygdala. And of course science helps in particularly dramatic fashion. If anxiety is our response to mystery, then discovering the truth and eliminating the mystery not only gives us more information about the world, it helps to soothe us. *Calm down there, amygdala. And trust in science.* The problem is that we often lack real answers. At that point, our own psychology can't help but get in the way. The anxiety we feel at confronting a real mystery encourages us to supplant the unknown with an answer that fits our preexisting worldview. For mystics, that means injecting God or some similar force into all the explanatory gaps; for materialists, it means maintaining faith that some prosaic explanation, far from mysticism, will ultimately emerge.

Science is often typified as a perfect, self-correcting system that compensates for our faulty wetware by gathering and totaling up evidence. Mystics, with their reliance on subjective experience, can't make the same claim. I find this standard analysis to be devastatingly accurate—to a point. My critique is that while we might look over the long haul and see a "perfect," self-correcting system, we aren't looking at today from the perspective of a decade, a generation, or a century from now. This means we might be looking at a completely laughable model of reality and calling it the most likely one—just because this is the time we live in and this is the best information our science has yielded to date.

Modern neuroscience provides a fantastic example of this: scientists are examining brain function at the level of the neuron to try to explain how consciousness is produced. Yet no one has figured out why neuronal firing and the interactions of neurochemicals produce thought, your feelings of being you, with your specific set of wishes and wants and fears, who feels a particular sensation upon perceiving the color red or enjoying the flavor of a good steak. There is a whole level of activity going on below the level of the neuron—a level we can't investigate so well because we don't yet have the instruments to do the job properly. And so, modern neuroscientists, for all their advancements, could prove to be just like the drunk looking for his misplaced car keys under the lamppost.

Where, we ask the drunk, *did you last see your keys?*

About three blocks away, he replies, a little man on his hands and knees.

Then, why are you looking here?

Because, he replies, *this is where the light is.*

In short, I'm not so much critiquing science as calling for more of it, particularly in areas derided as paranormal. The placebo effect was entirely disregarded for decades, a blip in the system, until the past ten or fifteen years. Now we know the placebo effect seems to depend on everything from the bedside manner of the doctor to the color of the pill. We're learning to manage belief so it works for us. But for years, the idea that what people believed could affect their health was just too far outside mainstream scientific thinking to be embraced as a field of study. Why would this be so? Thomas Kuhn's *The Structure of Scientific Revolutions* remains a kind of bible for those who would seek to understand science not just in its idealized form, but as it's actually practiced. The act of rigorously gathering and methodically analyzing data isn't carried out by idealized beings, after all, but people. And these people get invested in their findings and the theories they've signed their names to, and they protect the dominant way of looking at things—the current worldview science has built. So, at times, we not only have to wait for technology to catch up, but literally, for fallible scien*tists* to accept all the data that sci*ence* has amassed.

Science will eventually yield up all kinds of delicious truth—over the decades and centuries. The thing is, we're alive right now—not millennia from now—and so we have, as we've always had, a science that is doing the best it can with today's knowledge and technology base. The perfect aspect of science is its method, not the given set of information it's yielded to date. That store of information will change, and if history is any guide, it will continue to change dramatically—in ways we haven't even dreamt of. Look at the way science has dealt with quantum mechanics. By now, most people are familiar with the basic weirdness of the quantum universe, where particles regularly perform impossible feats: appearing in two places at once; communicating information across distances; blinking out of existence in one spot and reappearing suddenly in another.

Those strange subatomic operations underpin our every day reality, mainstream physicists long maintained, but do not manifest themselves *up here* at the macro-scale. In fact, in the realm of the paranormal, skeptics have long laughed at the way believers resort to quantum explanations for everything from the afterlife to consciousness and telepathy: *Quantum phenomena are so small and require such cold and stable environments,* they chortled, *that they could never persist long enough to be of any real importance in the operations of warm, wet biological systems.* But just in the past five years biologists have been discovering possible interactions between the micro- and macro-scales—interactions thought impossible till we found them. I discuss a few of these in chapter 3, but I am arguing that, given this state of affairs, we should be alert for and wary of dogmatism of any kind—be it religious or scientific. And I am further arguing that, before the evidence is in, many of us presuppose the answer that will best fit our worldviews—and soothe our overheated amygdalas.

What this means, in practice, is that professional skeptics deny any thought that validation for mysticism might arise from the quantum realm, while many modern believers go on seeing the quantum as the heavenly land we'll journey to when we die. I'd like to see these two sides in the debate start collaborating—or at least start taking each other seriously, and occasionally, there are signs such a thing might be possible. I'm a fan of the skeptic Brian Dunning, an atheist without an attitude, who runs a podcast on critical thinking called *Skeptoid*, which I highly recommend. And I'm perhaps even a bigger fan of the semi-retired entrepreneur Alex Tsakiris, who runs a podcast called *Skeptiko*. Tsakiris doggedly attempts to bring proponents of the paranormal and skeptics together for productive conversations—and sometimes he even succeeds. The brightest spot on the horizon, however, might be David Eagleman, a neuroscientist and author who has become the de facto leader of a new way of looking at the world, dubbed *possibilianism*. The creed of the possibilian is, I think, best summed up by Eagleman himself in an interview with the *New York Times:* "Our ignorance of the cosmos is too vast to commit to atheism, and yet we know too much to commit to a particular religion. A third position, agnosticism, is often an uninteresting stance in which a person simply questions whether his traditional religious story (say, a man with a beard on a cloud) is true or not true. But with Possibilianism I'm hoping to define a new position—one that emphasizes the exploration of new, unconsidered possibilities. Possibilianism is comfortable holding multiple ideas in mind; it is not interested in committing to any particular story."

In most respects, I applaud Eagleman. We live in a world of false certainties, a world in which a fundamentalist minister like Pat Robertson claims to know God, and Richard Dawkins claims with near-equal certainty and no less passion that no such God exists. The media, of which I am a member, foments this kind of debate all the time, in which only two polarized views are presented. We suffer through this in political coverage, too, listening to the most strident Republicans and Democrats and no one with an alternative point of view. And I think, as a people, our grip on reality itself is diminished. We are always being presented with binary choices, when reality is far more complex.

This book isn't going to provide a lesson in epistemology, the study of knowledge, which has gone on for millennia. It'd take a lot more space than I have to do that. But this is a book that asks the reader to recognize that there is a difference between knowledge and belief, and the bar for what constitutes true knowledge is set awfully high—far higher than we can attain throughout our society. So in court, for instance, we rely on eyewitness testimony when we also know, by scientific study, that eyewitness testimony is shockingly unreliable. What this means is that, as human beings, we traffic largely in belief. I think this fact could set us free if we let it. In not only admitting we don't know but acting on it, we open a door to conversation—

as opposed to debate—and the exploration of new ideas, a good-faith sifting through of the facts we have. I think, theoretically, most religious people can at least grope their way toward accepting this: in theological terms, doubt is often seen as a necessary part of real faith. Skeptics might have a harder time, because they usually profess that they deal only in facts. But as we'll see throughout this book, the arch-skeptic is as capable of seeing things according to his or her biases as the believer.

The result is that we don't merely live in a world of false certainties; we live in a world in which people at either extreme try and get those of us in the middle to buy into their particular fairytale version of reality.

This line of argument is normally waged solely against believers—the "old man with a beard" who takes away the sins of the world and greets us all in heaven with a sweetie. Psychologists also often talk in terms of the emotional or real-world payoff people receive in exchange for what we do—from the actions we take to the beliefs we hold. And for a long time, this sort of transactional aspect of belief was most evident in, well, believers. Those who believe in God or even a Godless afterlife have long been examined in terms of the benefit they receive for holding that belief: faith in the paranormal as a panacea for knowing death awaits us all, for instance. But if the New Atheists have succeeded in anything, it is in crafting a materialist fairy tale. Known as the Four Horsemen of the Apocalypse, Richard Dawkins, Daniel Dennett, Sam Harris, and Christopher Hitchens have created a similar payoff for believing there is no God. In their view, the only immortality available to us is the legacy we leave behind; and because they are right they will be remembered and judged well by history. And here on Earth, while they're alive, they get to feel smart while stupid goes on doing as stupid does.

Religion tries to scare people into believing its tenets. *Follow our rules or burn in hell.* Dawkins declares a kind of intellectual *fatwa* against belief itself, swaying fence sitters to his position through fear of ridicule. Forswear belief or be called a superstitious dullard, a dangerous fool? Atheists shall be known as "brights," and believers "dulls." Is this an improvement? It all works out the same. Believers and unbelievers alike operating as mean, petty bullies. It also betrays a startling ignorance of human psychology. When people are attacked, they become defensive, stop listening, and cling to their views more aggressively. In this respect, Richard Dawkins isn't fighting fundamentalism. He's calcifying it. And so the New Atheists have succeeded in pointing out the sandy foundation of dogmatic religious belief; they have lent succor and courage to the secular humanists too long confined to our cultural closet. But their cartoonish one-liners have also brought the same sense of polarized opposites to our discussion of faith and spirituality that already dominates and demeans our politics.

In the end, then, I'm not arguing for or against religion or atheism. What I'm trying to do is illuminate their common ground. Each side claims to have worked out a way of looking at the world that holds the ultimate claim on truth. Neither side seems likely to change its position. So in light of this, it seems we had all better do one thing in particular: learn to get along. I'm arguing that we learn to talk about so-called paranormal issues productively, so that believers and disbelievers alike gain a better understanding not only of how the world works but of themselves and each other. The way I see it, we're all land-based mammals on a planet with a greater surface area devoted to things that swim. We are all trapped on this same unforgiving rock, floating through space, with no rulebook for living other than the one we discover and write together. Under such circumstances, are we better off approaching each other in a posture of debate—or conversation?

I'm not alone in thinking this way. I should point out that among the New Atheists, Sam Harris seems to clearly understand the difference between a paranormal claim, or ideas related to spirituality (for lack of a better term), and supernatural propositions, or the kind of thinking codified into a religion. "The question of what happens after death (if anything) is a question about the relationship between consciousness and the physical world," writes Harris. "It is true that many atheists are convinced that we know what this relationship is, and that it is one of absolute dependence of the one upon the other. Those who have read the last chapters of *The End of Faith* know that I am not convinced of this. While I spend a fair amount of time thinking about the brain (as I am finishing my doctorate in neuroscience), I do not think that the utter reducibility of consciousness to matter has been established. It may be that the very concepts of mind and matter are fundamentally misleading us."

I suspect that most of us are reasonable enough to realize that systems of thought, whether religious or scientific, that have survived for centuries and for millennia must necessarily contain truths that are ours for the taking. What gets too little play, at least in our public discourse, is any sort of middle or integrated view in which both political parties have valid points to make, or both rationalists and mystics have something to teach.

But the good news is that we, as a society, are already beyond both Pat Robertson and Richard Dawkins. Or, at least, a world beyond these partisan yelling matches is *available* to us. Whether it is Eagleman, philosophers like Jean Gebser or Ken Wilber, or for that matter the Dalai Lama, more and more serious thinkers are recognizing that the most enlightened view allows for a rich dialogue between science and religion—not the dominance of one at the expense of the other.

Bringing more people into this kind of collaborative worldview won't be easy. And the media is one of the obstacles we need to overcome. Journalists often portray the fight between mysticism and materialism in stark terms—and through the lens of some dramatically phrased question, like, *Can science and spirituality coexist?* In the following pages, this book is essentially making one argument: That in the fullness of time, it's not only apparent that science and spirituality can coexist, as they have coexisted for centuries. The lesson is even more dramatic than that. The lesson is that they can serve one another.

We are about to embark on a tour through a series of fundamental questions about the nature of human existence. In a sense, we will all be like the man, Paul, we saw at the beginning of this introduction. You may hear the footfalls and bangings of various ghosts—or the workings of human imagination.

You will definitely encounter a UFO.

Mind reading, a trip to the moon, and spiritual ecstasy are also all on the agenda—and we'll get a few glimpses of potential afterlives, besides.

The answers, when they are available, sometimes sacrifice the specificity of a false certainty for the accuracy of the unknown, as they did during my time with Lou Gentile. Paul and his family reported a wide range of phenomena—including doors that rattled in the middle of the night and drawers that inexplicably shook in their dressers, all in the wake of a death in the family. But the most interesting thing was that series of photos Paul showed us.

I'd seen lots of photos like these over the years, and Gentile had probably seen thousands of them. Paul seemed certain that each comprised evidence that a ghost—maybe the ghost of his father—was floating around his house. But light reflecting from nearby, out-of-focus dust particles, can create a halo that turns a dust mote into what some take to be evidence of spirits. So Gentile, slowly, gently, and deliberately, explained to Paul how dust floating in the room might have caused the vast majority of his strange images.

When we left, I took this as an occasion to talk to Gentile about how easily people delude themselves into thinking every photographic anomaly is a ghost. Lou agreed. But he always tried to counter my skepticism with examples of phenomena that couldn't be so easily explained away. And so, on one of my visits with him, he set up a projector screen in his home and told me a story about what he considered perhaps the best photo he ever took in all his years investigating haunted houses. "I want you to see a photo," he said, "that isn't just dust."

He was looking into a reported haunting in Connecticut, he explained, and he was standing in an upstairs hallway, just waiting for something to happen, when a sudden movement flickered in his peripheral vision. Turning quickly, he saw it—a long shaft of light, slowly undulating toward him, like a snake floating in the air. Gentile had a camera hanging around his neck and recovered his senses just in time to snap a single photo as it passed him by.

Gentile told me this story many years after the event in question occurred. The projector screen behind him was still blank. Then he clicked a button on the projector and the photo he took appeared before me. The light he claimed to see was in the image, maybe six feet long, its form clear and consistent, its leading edge caught in the middle of a dip. I stayed focused on the light itself, but then Lou showed me why he thought this light was so intriguing. And there it was, in the lower corner of the photo, a shadow, seemingly cast by this odd shaft of light—suggesting that *some*thing, however unlikely, was actually there.

"What do you think that was?" Gentile asked, to which I could only reply, "I don't know."

In the coming pages, these interrelated tales will force us to use that same dirty, three-word phrase—"I don't know"—on a regular basis, heating up our emotions in the process. But I think a tour through our own fractured universe, in a sense our own fractured selves, is exactly what we need right now.

1

On Death and Not Dying

The Provocative Beginning of Near-Death-Experience Research

Our nada who art in nada, nada be thy name.

—Ernest Hemingway, "A Clean, Well-Lighted Place"

I can hear my mother wailing, and a whole lot of scraping of chairs. I don't know what it is, but there's definitely something going on upstairs.

—Nick Cave, "Dig, Lazarus, Dig!"

She sipped tea with the rhythm of an addict and ate chocolates the same way. Swiss chocolates, from her homeland, her fingers steadily working, her face going momentarily serene whenever cocoa met tastebud. These small pleasures were essentially all that was left her. The great lady's world had shrunk along with her tiny frame. She was one of the most significant figures of the twentieth century, shaking hands with dignitaries the world over, accepting not their congratulations but their thanks.

All that was over now. The great lady had diminished with age, like a paper going yellow in the tick of time. She was seventy-eight years old but seemed even older, her skin wrinkled in deep folds, her voice and body weak from a series of strokes that did her in over a period of years.

She knew.

She had seen all this before.

Up close.

But this time, the looming presence in the corner of the room was there for her. Elisabeth Kübler-Ross was dying, in an Arizona nursing home, and still, complete strangers came to her. They had been coming for years, people who had been touched by her work in some way and wanted, sometimes with feverish intensity, to meet her. Once, as her friend Fern Welch made her way inside her tiny private room, she found Kübler-Ross beset by two young women, who she remembers literally kneeling on the floor and bowing before Kübler-Ross. Welch made her way around the two prostrate girls to give Kübler-Ross a customary hello kiss. But when she leaned in, the old woman whispered to her, "Please, get them out of here!"

Welch had seen all manner of Kübler-Ross's fans show up over the years, and shooed the girls away. But this particular episode stayed with her. The irony was so great. Two young girls, venerating an old woman and stealing a portion of the little time and energy she had left. Kübler-Ross died not so long after this in August 2004, culminating a long, public descent.

Most people know Elisabeth Kübler-Ross as the writer and psychologist who wrote *On Death and Dying*, which remains one of the most important books of the twentieth century. Published in 1969, her tome's message was brutally simple: We, as a society, treat our sick and dying loved ones as something less than human. Hospitals relegate terminally ill patients to back rooms, where the specter of illness can be kept out of sight. They enforce short, strict visiting hours, as if to make sure the dying suffer alone. And they strip the sick of their dignity, failing in most cases even to acknowledge the patient's terminal diagnosis. This description, written in the present tense, appears jarring and somewhat inaccurate today, some forty years later. But at that time Kübler-Ross merely described the world as it stood. She had gone into a hospital to help the sick and found that everyone was ill—that seemingly healthy family members, doctors, and nurses were all victims of the human condition.

Doctors saw the death of their patients as a professional failure, so the terminal diagnosis went unspoken; the nurses felt trapped by this, darting in and out of the room and avoiding eye contact, the better not to give away the truth; and loved ones struggled with what to say. Our symptoms may differ, but the underlying distress is universal. Because, Kübler-Ross wrote, every last person coming into contact with the terminally ill patient is reminded of his or her own mortality. What this means is that we love the dying, pity them, mourn for them, wish to heal them—and hate them, too, for reminding us that one day we also will lie there helpless, flat on our backs till the end. This cocktail of emotions is so potent that Westernized societies, deceived by the seeming omnipotence of modern medicine, tucked the dying out of sight, hiding them behind closed doors, beeping machines, and IV stands—a veil of technology that soothed everyone but the terminally ill.

Kübler-Ross became most famous for outlining five stages of grief—denial, anger, bargaining, depression, and acceptance—stages that subsequent academics have since criticized and amended in various ways. But her real accomplishment was in getting cultures all over the world to openly discuss death—the reality we'd all been busy hiding from. She accomplished this feat in a slim 276 pages. And in so doing, she became the public face of a then-new hospice movement.

It is impossible to quantify the impact Kübler-Ross's *On Death and Dying* had on the world at large, but by any account it was massive. The emotional maelstrom unleashed in the wake of her book literally remade end-of-life care. And today, laypeople still turn the pages of *On Death and Dying* when there seems nowhere left to go. Medical programs still list her as required reading. And in brief newspaper and magazine summations of Kübler-Ross's life, her story often seems to end there, in about 1974, on the heels of her greatest professional achievement. But she lived for roughly thirty more years. And for our purposes, the real entry point into her life story is the subsequent turn she took into the paranormal. In 1975, in fact, she put her then impeccable reputation on the line by penning the forward to Raymond Moody's *Life After Life*—literally the first book ever written about near-death experiences (NDEs).

By now, the lore of the near-death experience is well known: The tunnel, the light, the life review, the reunion with loved ones. But in 1975 the term NDE had never previously been used, so the press and academia both were shocked at Kübler-Ross's involvement. The story was clear enough for every reporter to see: the brave lady who asked us not to shrink from the reality of death suddenly suggesting death may not exist at all. There is much to be learned from this twist in the life story of Elisabeth Kübler-Ross. Perhaps most important, venturing into these realms is dangerous—to our reputations and relationships, to our filter for what's real and what's fiction.

By 1980, in fact, the same woman who cried "Bullshit" about Western medicine had succumbed to bullshit of another flavor: she was consulting mediums who claimed to channel spirits of deceased loved ones and "higher entities." She hosted New Age–themed spiritual retreats for widows. Her husband left her and became the main caretaker to their children. Her most prized medium proved fraudulent. Kübler-Ross's fall from grace became a running gag in the media, her life story ballooning into something too vast and complicated to be captured by any single narrative. Her son, Ken Ross, minds her legacy now—the foundation formed in her name. And he knows, up close, how hard it is to pull together a coherent story about his mother. He says HBO has spent years trying to develop a movie based on his mother's life. Numerous screenwriters have come and gone, none delivering a workable biography. "My mother's life was complicated," says Ross. "They told me they couldn't figure out which story to write."

But the Kübler-Ross story isn't only hard for screenwriters to encapsulate. It's hard for all human beings—or at least those of us with normal, functioning brains. We are, as a species, neurologically uncomfortable with ambiguity. Imaging studies of the human brain in action demonstrate that the fussy little onboard computers in our skulls send out anxiety messages when confronted by conflicting or confusing information. As a consequence, we have a natural, internal impetus to settle on an interpretation that removes any perceived conflict.

If we're not cognizant of our own biases, this probably means choosing an interpretation that preserves our current worldview and disregards contrary evidence, replacing the accuracy of *I don't know* with the false certainty that we do. So, how *do* we reconcile the brave, passionate, hardnosed pragmatist who taught us about death with the New Age queen who lost so much over her commitment to a psychic? Well, according to the people closest to Kübler-Ross, and just as modern neuroscience would predict, we don't. We see her as the great lady. Or we see her as a crank. We bow to her memory. Or we smirk at the mention of her name.

"People tend to see Elisabeth as they want to see her," says Rose Winters, a friend of Kübler-Ross. "It's hard for those of us who knew her. Because people don't acknowledge *all* of her. They don't see her as she really was."

We see her, it appears, in much the same way we see the paranormal (or the political), as if we only have two choices: to passionately embrace or hotly reject. But there is a messier, truer view, one we need to draw closer to if we are to understand her, or even ourselves, let alone the paranormal.

||||||||||||

ELISABETH KÜBLER-ROSS WAS BORN in Zurich, Switzerland, on July 8, 1926, the first to emerge among triplets. She weighed just two pounds and was not expected to survive. As a child, perhaps mindful of her own early frailty, the young Elisabeth Kübler nursed any injured animal she found, including a crow she fed and protected till it was strong enough to fly away. She defended weaker kids from schoolyard bullies. And she even bounced a book of psalms off the head of a preacher who had unfairly punished one of her sisters. Though she was later typified as a New Age faerie queen, the truth is she was a bit more like Keith Richards—a rebel by any accounting.

She first rebelled against her father, an assistant director of Zurich's biggest office supply company. "He had dark brown eyes that saw only two possibilities in life," his daughter would later write. "His way and the wrong way."

In the chauvinistic Switzerland of the early 1940s, her dream of being a doctor was considered just that. And one night her father sat her down to talk about her future. She was so responsible, he said, so capable, he thought she would make a fine . . . *secretary*.

"You will work in my office," he told her. But the thought of being stuck in his boring office, following his boring orders, and furthering the aims of the boring office supply industry, rather than doctoring, made Elisabeth Kübler half-nuts. "No, thank you!" she told him.

His counteroffer? "Then find work," he said, "as a maid."

Having no other means to support herself, she did just that. And in the ensuing years, she made her own way in the world. She left home, attended medical school, and met her husband—an American med student named Manny Ross. Her gender shaped her path. After she moved with Manny to America and became pregnant, the only residency program that would have her was the one she didn't want: psychiatry.

Still, they needed the money. So she took a position at the Manhattan State Mental Hospital, working in a small unit with schizophrenic women. The head nurse allowed her cats to freely roam the ward, pissing and defecating among the patients. The entire asylum carried the ammonia stink of cat urine. And patients were punished for showing signs of their mental illness—beaten with sticks, subjected to electroshock treatments, and experimented upon with drugs like LSD and mescaline. "What did I know about psychiatry?" Kübler-Ross later wrote. "Nothing. But I knew about life and I opened myself up to the misery, loneliness and fear these patients felt. If they talked to me, I talked back. If they shared their feelings, I listened and replied."

She was already opening herself up to the role she would play to the dying: the woman who shared their burdens and received their woes. But psychiatry never felt right to her until she and Manny moved to Denver for hospital positions. There, a colleague asked her to fill in and deliver a two-hour lecture he couldn't make. She cast around for a subject. She hunkered down in the library. She walked the hospital halls, wondering what topic would be suitable for a general audience of medical students and residents in various specialties. The answer came to her at home, as she stared into a pile of dying leaves, rake in hand. At the time, in 1964, death was not really a hot topic in medicine. In fact, when Kübler-Ross went back to the library to see what was available on the psychology of dying she found precious little: a single, dense, academic psychoanalytic treatise; some sociological studies on death rituals across cultures. She realized she would need to do her own research. But for her talk, she spoke only for the first hour. Then, during the break, she retrieved a patient she met in the hospital's wards: a sixteen-year-old girl who was dying of leukemia. When the students returned, Kübler-Ross explained the girl's terminal condition and opened the floor to questions. No one raised a hand. So she called on students, requiring them to come to the stage and *think* of a question.

These were med students. They asked about the girl's blood count, the size of her liver, her chemotherapy trials. The girl grew furious and began talking, unbidden, about what it was like to be sixteen and given only a few weeks to live; what it was like to never go on a date or have a husband; and how she was coping with it all. When she was finally wheeled from the room, the audience sat in heavy, dumbfounded silence. And gently, in her soft, Swiss accent, Kübler-Ross diagnosed what troubled them. *Your reaction is a product of your own mortality*, she told them, *which the girl forced you to confront*.

In this sense, they had not been looking at a sixteen-year-old girl at all. They had been looking into a mirror. The experience was so powerful that Kübler-Ross stopped questioning her own commitment to psychiatry. And when she and Manny subsequently moved to Chicago, she took a position at Billings Hospital, which was affiliated with the University of Chicago, and began her mission: to reconcile the world of the dying with that of the living. She grew famous for her efforts. But what is less well known is that during her years in Chicago many strange things happened to Elisabeth Kübler-Ross. And she also found an unlikely professional companion.

The Reverend Mwalimu Imara (then named Renford Gaines) was assigned to Kübler-Ross by the hospital's administration, almost as a kind of bodyguard. No one thought she would be the victim of actual violence. But the academic seminars she began on the topic of death and dying caused great controversy in the hospital's halls, so Imara, one of the hospital's chaplains, walked alongside Kübler-Ross as a sign she was not alone. She had the administration's support. The truth is, Imara wasn't that experienced himself yet, certainly not in the duties he'd be attending to beside Kübler-Ross. And he watched as her colleagues lied to her, again and again. "I am here," she would say, "to meet with your dying patients."

"No one here," she was told, "is dying."

No doubt, they thought they were doing the right thing. They thought it better for the patient not to discuss what was happening. No matter. She could read a patient's chart like any other doctor and found the terminal for herself. Imara still remembers watching Kübler-Ross attend the first patient they ever visited together. The woman sat alone in the dark, perched on the edge of her bed. Uneaten food rotted on a stack of trays left on a nearby table. Kübler-Ross pulled up a chair and sat down across from the woman. "And how are things going for you?" she asked.

The patient, her head down the entire time, finally looked up at Kübler-Ross. "I'm hungry," she said.

Kübler-Ross stood, opened the blinds, and called the nursing staff down the hall. "Get this woman fed," she said. "Help her eat."

The next day, Kübler-Ross returned to the woman's room. The blinds were still open. The uneaten food had been thrown away. The woman looked fitter. Kübler-Ross sat down beside her on the edge of the bed, and the woman smiled. The two ladies sat like that for a long time, grinning at each other. "Now," Imara told me, "just multiply that moment by hundreds or thousands of other moments just like it."

For Imara, bearing witness to scenes like these transformed his position alongside Kübler-Ross from assigned functionary to more than willing collaborator.

The terminally ill were being neglected. They sat alone in their rooms waiting for the culmination of a death sentence that had never been formally pronounced. And as Imara puts it, "hurricane Elisabeth Kübler-Ross" helped them to go on living as best they could manage until they did die. This meant reconciling relationships, acknowledging their feelings, and finding what joys they could. Most of the hospital's professional staff allied themselves against these efforts. Some doctors and nurses accused Kübler-Ross of ghoulishness. One nurse asked the psychologist if she enjoyed telling a twenty-year-old man he was dying. The signs advertising her seminars were torn down. But Kübler-Ross seemed to gain strength from the opposition. She recognized the resistance she faced as a symptom of the illness she treated.

And together, she and Imara didn't just challenge the medical establishment. They sat by the bedsides of people who described incredible happenings: *I left my body, I floated up to the ceiling, I saw the doctors resuscitate me.* While their bodies lay below, in distress, they rose above them and felt an overwhelming sensation of peace. There was more, much more, and these patients wanted someone to tell them they weren't crazy. But at this point in her life, Kübler-Ross not only disbelieved in organized religion, she viewed death like most Western doctors. Death meant the end—the terminus of termini, the obliteration of all possible beginnings. Even Imara, the reverend, was unprepared for these near-death tales. Many if not most sects of Christianity accept the virgin birth and the resurrection of Jesus but disavow paranormal happenings in our time. And here they were, what Imara calls the "Mutt and Jeff team," hearing classic near-death experiences years before anyone had coined the phrase, years before the phenomenon was widely known. These stories suggested death was merely a gateway from one existence to another, from one incarnation to a newer, more profound one.

The stories told by children were often the most incredible—and consistent. "So many kids would start telling these stories, in the days before they died, about spirits visiting them," said Imara. "Some drew pictures recording the date and time of their deaths."

One girl told Kübler-Ross she had withheld an NDE from her mother because "I don't want to tell mommy there is a nicer home than ours."

Kübler-Ross sat by the bedside of one boy who was dying in the aftermath of an accident. "Everything is all right now," the boy told her. "Peter and my mother are already waiting for me."

Kübler-Ross knew the boy's mother was dead, but thought his brother, Peter, was still alive. Peter had suffered serious burns in the accident and been taken to another hospital. She left the room about ten minutes later and was stopped on her way past the nurses' station. There was a phone call for her, from a nurse at the hospital where Peter had been taken. "Dr. Ross," the nurse said, "we just wanted to tell you that Peter died ten minutes ago."

To the skeptically minded, this is all meaningless—an anecdote captured under noncontrolled conditions and interpreted according to the belief that such a thing as life after death is even remotely possible. But Kübler-Ross was no pie-eyed believer. Her only real concern, like that of most people, was with what she saw in front of her—not the dead, but the dying. "We weren't looking for this," says Imara. "It was just happening, again and again, *to* us."

Once, Imara looked in, by himself, on a little girl dying of leukemia. She had been visited, she said, by a man she didn't recognize. He wasn't a doctor. He had been dressed in civilian clothes. Imara did his best not to upset the girl. But inside he felt anxious. Somehow, hospital security had lapsed. Somehow, a stranger had spent time alone with an exceedingly vulnerable little girl. Imara stayed near the girl's room until her parents arrived, then intercepted them before they could go inside. "Your daughter is fine," he told them. "But there is something I'd like to talk to you about."

With Kübler-Ross accompanying him, he led the parents into a large conference room. "The hospital will make sure it doesn't happen again," he said, "but someone came in to see your daughter. A man. He didn't hurt her. He was apparently very kind. Hospital staff will be on the lookout."

Her parents looked understandably upset, until Imara relayed the girl's description of her visitor. Then, the girl's mother got . . . *interested*. The description matched that of her own brother, the girl's uncle, who had died before the girl was born. Even the description of his clothing matched what her uncle had been wearing at the time of his death. The scene quickly shifted to the little girl's room, where her mother had her repeat her story. Imara says that, after multiple strange happenings like these, he and Kübler-Ross agreed to start taking notes not just on the psychological impact of death and dying but the weird stuff, too. "In her office," he said, "she filled a couple of deep filing cabinet drawers just with stories like these. And when we could, we corroborated them."

One of the most intriguing stories was that of Mrs. Schwartz, who appears and reappears rather dramatically in the tale of Elisabeth Kübler-Ross. A middle-aged mother with Hodgkin's disease, Schwartz went into cardiac arrest in the hospital as she was being wheeled off an elevator. Imara was among the people who witnessed the scene. Her clothes were stripped off to give doctors access to her body. A med student stood by and took notes, a dispassionate act Imara says was designed so early med students could have something to do when real doctoring was necessary.

Doctors successfully revived Schwartz. And she later told Kübler-Ross and Imara how the scene had looked to her, from the position she assumed after her heart stopped—out of her body, up near the ceiling. She accurately described the resuscitation efforts and comments made by the people at the scene. She claimed she floated behind the med student and even looked at his notes. He had drawn doodles at the top of the page, she said, and she described those, too.

Skeptical, Imara retrieved the student's notebook and looked for himself. Mrs. Schwartz's description matched—right down to the doodles.

Imara says Kübler-Ross was "knocked from her moorings" by these strange events. And all the ordinary explanations they could dream up couldn't account for all of the weirdness.

As time passed, word had spread about Elisabeth Kübler-Ross's death and dying seminars. *Life* magazine printed an extensive profile of her. Book publishers came calling. And Kübler-Ross began work on the book that would become *On Death and Dying.* By this time, Kübler-Ross had accumulated enough stories about NDEs and deathbed visitations that she was giving serious consideration to publishing them along with her famous psychological stages.

In fact, according to Imara, in its original incarnation, Kübler-Ross wrote a concluding chapter to *On Death and Dying* in which she detailed numerous near-death experiences. She wrote that her research had suggested something extraordinary, comforting—and beautiful: at the very end, after the tremendous emotional and physical pain associated with death, there may be another life to live, another plane of existence to visit.

As might be expected, given the closeness of their professional relationship, she asked Imara to give the manuscript one last look. Manny and the children were asleep, so it was just the two of them up late at night in her house, going over manuscript pages, patching in quotes, and tweaking sentences. But the most important conversation of the evening was about the last chapter.

The room was filled with smoke from Kübler-Ross's cigarette addiction, Imara remembers. And they had long since moved on from tea to whiskey sours. "Do I put this chapter in?" asked Kübler-Ross.

"Not if you want it published," replied Imara. "You don't tell them about the kid on the second floor who spoke to his dead sister. You don't talk about getting into someone's house, after they died, and finding a picture he drew with a map and a clock, of the exact time and place of his death. You don't do it. That shit will not fly."

Kübler-Ross was already taking on an incredibly taboo topic: death. The idea that she might end a book that already challenged the biases of the powerful medical profession with an *oh by the way* chapter, suggesting the possibility of an afterlife, would have been way, way too much. Her entire life's work would have been dismissed. And in her last chapter she would have given people the means with which to attack her. So fraught is our relationship to the paranormal that Kübler-Ross was forced to consider self-censorship.

The house was quiet, and so were they. This was a pivotal discussion, and both of them knew it. Elisabeth Kübler-Ross was not someone who routinely cared what cultural or political forces were at play. She only cared about what she took to be the truth. But in this case, the stakes were so high, the suffering she had seen among her patients so great, she agreed. She would hang on to this chapter. She would put these tales of the dying on a shelf.

That night was perhaps the most crucial of Kübler-Ross's professional life, and in the lives of millions of people with terminally ill family members all over the world whose lives would be changed for the better by her book. Because that night Elisabeth Kübler-Ross did what was, in this instance, the right thing. She consigned the paranormal to the dark.

||||||||||||

THE NEAR-DEATH EXPERIENCE IS a phenomenon associated with the paranormal, but it is the product, in its modern incarnation, of science. It is no coincidence that Elisabeth Kübler-Ross started hearing these remarkable stories so often in the mid-1960s. It was in 1963 that CPR was first widely publicized and adopted, allowing doctors to save more critical patients than ever. The unexpected result was a boom in reports of an experience that had been rare but nonetheless described for millennia.

The Greco-Roman historian Plutarch related the story of Aridaeus of Soli, who "died" after a fall and came back with a story to tell. He reported traveling to other realms and meeting with a younger relative who had already died. His personality changed so much after his NDE, and so much for the better, that his name was changed to Thespesius, which meant "divine" or "wonderful." This is a particularly interesting detail, given that modern day NDErs, as they're known, routinely change their lifestyles for the better.

Thomas De Quincey included the following account in *Confessions of an English Opium Eater*: "I was once told by a near relative of mine that, having in her childhood fallen into a river, and being on the very verge of death but for the assistance which reached her at the last critical moment, she saw in a moment her whole life, in its minutest incidents, arrayed before her simultaneously as in a mirror; and she had a faculty developed as suddenly for comprehending the whole and every part."

In the 1887 book *Euthanasia or Medical treatment in the Aid of Easy Dying,* Dr. William Munk quotes an Admiral Beaufort of the British Navy: "The whole period of my existence seemed to be placed before me in a kind of panoramic review, and each act of it seemed to be accompanied by a consciousness of right or wrong, or by some reflection on its cause or its consequences; indeed, many trifling events which had been long forgotten, then crowded into my imagination, and with the character of recent familiarity."

There are several reasons these accounts, and the NDE, have gained a kind of traction among scientists. Number one, the number of people reporting the phenomena is impressive. Studies show that roughly 6 to 12 percent of cardiac arrest victims report NDEs, and though particulars vary wildly, the general character of the NDE—the arc of the narrative—remains the same across accounts. Radiation oncologist Dr. Jeffrey Long gathered more than 1,600 NDEs, which he analyzed in a bestselling book. He identifies twelve recurring elements of the NDE, including the sensation of leaving the body, heightened sensory ability, intense and generally positive emotions, encountering a brilliant light, mystical beings and/or deceased loved ones, passing through a tunnel and visiting other realms, and undergoing a life review. These are commonly discussed in media accounts, but perhaps most intriguing, many NDErs also report feeling that time and space as we know them have ceased to exist. Thus, their life review seems to take place in an instant, yet they are aware of each event as it flashes before their eyes.

Long's breakdown of the NDE seems most relevant, for our purposes, because it is the newest and arguably the most comprehensive and because his results don't differ markedly from most other surveys. The relative consistency across accounts is usually regarded, among believers, as indicative that one central phenomenon is at work here. Skeptics, of course, focus on the differences among people's stories. But there are two things I find compelling: in our culture, the NDE itself is an orphan—entirely unwanted by those at the extreme poles of belief and unbelief. Skeptics are inclined against the NDE representing any objective reality. Fundamentalist Christians often deny it any veracity on the grounds that it presents the possibility of an afterlife steeped in a nondogmatic, irreligious spirituality.

The problem, for those of us in the middle, is that these polarized views tend to obscure what might be most important about the NDE: its impact. In sum, whatever happens in an NDE, experiencers tend to *react* to it in much the same way. The vast majority take the experience to be real, and they change. They more greatly value their families and friends. They become less materialistic. Most remarkably, they no longer fear death.

In other words, NDE research confirms what Kübler-Ross experienced in her own journey through the halls at the Chicago hospital: people who undergo some kind of near-death experience feel a sense of contentment. For them, death looms as a kind of spiritual eject button, separating them from their living friends and families yet popping them out into an infinitely more expansive existence.

This last fact lends the NDE real heft. If this is an illusion, it is one that is irrefutably important to nearly everyone who experiences it. This also gives psychologists something real and measurable to work with. The historical accounts provide further verification, corroborating the NDE across centuries and millennia as an experience endemic to humans—and not borne from the power of suggestion.

I reached out to a former military nurse, a retired colonel named Diane Corcoran. She first encountered the NDE phenomenon in the late 1960s and early '70s, in Vietnam soldiers, well before Raymond Moody's bestseller. Corcoran is career Army, and in our conversation she spoke with the matter-of-fact authority of a lifelong member of the military. In 1969, while serving in Vietnam, she tried, like Kübler-Ross, to make herself a receptive ear to all the wounded soldiers on her ward. But one day, a soldier told her about the accident that left his body wrecked. He had been knocked unconscious, yet he remembered it all. Because after he was struck by a moving vehicle, he left his body and looked down on himself. He saw other soldiers come running to help him, or his body, anyway, which lay still far beneath him. Then he saw a tunnel nearby, floated toward it, and peered down its length. He hovered there, seemingly between two worlds, certain that whatever happened to the husk of him below, he was going to be all right. Corcoran listened to the soldier and blinked dumbly. "At first I said nothing," she remembers. "It seemed like the smartest thing to do. I mean, I just didn't have any idea what he was talking about."

But then she understood something else: the soldier was at once convinced of the reality of what he had undergone and worried that other people would think he was crazy. He needed her validation. He needed her to honor the importance of what he had gone through. And so she said, with supreme understatement, "That must have been an incredible experience."

The skepticism these early experiencers encountered suppressed any widespread public knowledge of the phenomenon for quite a while—at least six years that we know of, by my count, given that Kübler-Ross herself wrote a chapter on the subject before deciding it best served the world gathering dust. It is, of course, impossible to determine the number of people who decided to shut their mouths and keep the white light to themselves. But it's not hard to understand why. Debates on the paranormal often break down in rancor, which is surprising because in an epistemological sense this might be one riddle to which we can never really *know* the answer.

Death is, by definition, the end of life. Anyone who tells us about the death experience is very much alive, rendering their view of the afterlife, one way or the other, suspect. Further, materialist arguments can only be so authoritative on the subject. What happens to a person's consciousness after death can be pondered, imagined, or hypothesized, but not directly observed or measured. The thought that consciousness is purely a product or epiphenomenon of the brain is the mainstream scientific view. But the truth is, as we'll see in chapter 3, the source and nature of consciousness remains an unsolved mystery. There are numerous aspects of the NDE that can be studied by science, but what happens when we die still seems a subject most closely associated with philosophy.

This strikes me as rather obvious. I only bring it up because so many people seem to forget it. And I believe that if we remember our own inability to deliver a definitive vision of what lies beyond death, we might find ourselves a little less threatened by others' opinions. Welcome to reality: you don't know what happens when you die, and the person arguing with you doesn't know either. Some people don't seem to care for an afterlife. Me? I'd like to believe in an afterlife. My mother has died; so have my oldest brother and a dear brother-in-law. But I'm also not interested in fooling myself.

My mother's death is, in part, what drew me into investigating Kübler-Ross. As my mother slowly succumbed to a long-term illness, nothing at all paranormal happened. But I did receive the benefit of Kübler-Ross's work, rereading *On Death and Dying* over a couple of difficult days. The story she tells of a farmer is the one that haunted me. The farmer she describes took his beloved wife to a big city hospital in the hope of saving her. He wanted to spend every minute with her. He wanted to sit beside her during the day and sleep with her at night. But he was not allowed to stay in the intensive care unit for more than five minutes an hour. He took what he could get, coming and going under orders, staring at his wife's white face and holding her hand till he was told to leave. In contrast, nearly forty years later, my family was allowed to stay with my mother everywhere, and at all times—when she was in a private room, and when she was in intensive care.

My father, married to my mother for fifty years, issued a simple order. "Your mother asked that she never be left alone," he said. "So that's it. We're not going to leave her alone."

And we didn't. She had a family member by her side in the hospital, twenty-four hours a day, seven days a week. My father logged the vast majority of hours with her. The nurses and doctors were so taken with his dedication that they eventually put my mother in a private room and made a bed next to her for my father to sleep in.

I spent comparatively few nights there—two per week. But one night, toward the end, a nurse's aide woke me. "I had to meet you," she said. "I've met the husband, and the daughters. Now I had to meet the son. Your family's dedication is amazing."

I think of the old man Kübler-Ross described who could not stay with his wife for any extended period of time and feel so tremendously grateful that at the end we were able to spend this time with my mother. On maybe the most emotional of those days, my sister Karen and I spent an hour or two sitting on either side of my mother's bed. My father had recently discovered the hospital television included an easy listening music channel. I can't remember what song was playing. But we sat on either side of my mother, each holding one of her hands. My mother sang a little. Her voice was weak. She asked Karen to sing for her. As Karen's voice rose, my mother closed her eyes. She smiled. "Isn't this cozy?" she said. "Isn't this nice? We'd never have done this if we were home."

She repeated that last fact a couple of times. And although her medication usually made her drop off to sleep within a few minutes, she stayed awake for nearly half an hour that night. "Don't fight," she said suddenly. "Life's too short. Don't ruin it by fighting."

Our family house, like so many, was not always peaceful—a subject my mother had never addressed. To have her mention it at all, saying no more words than necessary, released a tension built over many decades. She hung on to our hands. She fell asleep, smiling. This was the kind of moment Kübler-Ross sought to facilitate by encouraging medical staff and families and the sick to acknowledge the predicament they faced and settle their unfinished business.

Had Kübler-Ross not decided to focus her book purely on the end of life, rather than the afterlife, it's possible my family and countless others would have suffered far more difficult experiences than we did. Kübler-Ross may even have been fated to obscurity. That said, I think it's high time we—believers and unbelievers alike—acknowledge what we do know for sure about the NDE: right now, the ultimate conclusion we each choose to draw seems based more in the vision of the world we bring to the data rather than the data itself. Believers applaud the researchers who conclude there is an afterlife; skeptics celebrate those who decide the NDE is the product of brain function. I argue that, in many cases, the data merely become a means of landing ourselves in the world we most wish or expect to see.

By Jeffrey Long's count, at this time skeptics have put forth no less than twenty explanations for the NDE. The number, I think, has ventured so high because people report NDEs at times when they were physically dying, and times when they weren't dying; at times when they were merely in fear of death, and at times when they felt no fear at all; at times when they were under general anesthetic or some cocktail of drugs, and at times when they were completely unfettered by pharmaceuticals. Skeptics, then, are searching for some singular explanation or combination of explanations that can occur in various states of consciousness, yet each trigger an NDE.

It has been speculated numerous times that some medication administered by hospital staff helps produce the NDE. But thus far, no clear correlation has been found between the use of recreational or medicinal drugs and NDE incidents. Dr. Karl Jansen, a New Zealand-born psychiatrist, has argued for a connection between these mystical flights and the drug ketamine. But ketamine is an ass-kicker of a narcotic with profound dissociative effects; while it *can* lead to experiences similar to the NDE, it can also lead to myriad other sensations—trippy happenings, feelings of falling, even psychotic experiences. Intriguingly, though Jansen started as a skeptic, ascribing purely material theories to the phenomenon, he has come to believe both ketamine use and NDEs may represent a glimpse at a larger, more fundamental reality. But there seems to be some fundamental difference between the two trips, so ketamine, at this point, looms more as an idea than an explanation.

Dr. G. M. Woerlee has claimed that NDEs during resuscitation are produced by cardiac massage, which he says can produce enough blood flow to the brain to allow consciousness. But as we'll see in a moment, it seems unlikely that the amount of blood flow generated by CPR is sufficient to sustain consciousness at a high enough level for the brain to formulate the orderly, vivid experiences associated with an NDE. We can best view that, however, through the lens of the perhaps most publicized skeptical theory, which belongs to U.K. psychologist Susan Blackmore. She delivered the real poison pen letter for NDEs in her book *Dying to Live*; I focus on her version of events because most modern debunkings of the NDE originate with her "dying brain hypothesis."

Blackmore is an elegant writer who makes her case over the broad length of a book. But her argument can be summarized rather quickly: in short, a lack of oxygen, known medically as anoxia, results in a narrowing of vision, fading first at the edges, that would create the illusion of a tunnel and light. Further, endorphins released at the time of death, under stress, might serve as the source of euphoria associated with NDEs. Accurate perceptions made while "out of body" are created out of memory, expectation, and the sensory detail accumulated before consciousness was lost. A sensation of timelessness is created by our loss of a sense of self as the brain breaks down.

Dying to Live stands as the most comprehensive argument anyone has launched against the NDE as a window on the afterlife. That said, some of Blackmore's key points don't stand up well under scrutiny. A number of medical doctors have conducted their own research into NDEs. Sam Parnia, Michael Sabom, and Jeffrey Long have all written books on the subject, and Parnia, as do others, notes that many people who report NDEs aren't lacking for oxygen at the time. Perhaps even more damaging to Blackmore's case, in his book *What Happens When We Die?*, Parnia writes, "If the dying brain theory were correct, then I would expect that as the oxygen levels in patients' blood dropped, they would gradually develop the illusion of seeing a tunnel and/or a light. In practice, patients with low oxygen levels don't report seeing a light, a tunnel or any of the typical features of an NDE; and, in fact, this experience has never been reported by any other doctor or scientific study as a feature of a lack of oxygen."

While Blackmore focuses much of her attention on the subject's experiencing a lack of oxygen, her explanation falls apart most readily in that very instance (the same goes for Woerlee's cardiac massage theory). According to Parnia, "Memory loss is so closely associated with any insult to the brain, whether from a blow or a lack of oxygen, that the degree of memory loss is used as a diagnostic tool to assess the severity of the brain damage."

Such memory loss usually encompasses a period of time from minutes to days or even weeks before losing and after regaining consciousness. So even if people were regaining awareness as chest compressions were administered, they still seem deeply unlikely to report *anything* at all or at best fragmented memories. Yet people who undergo NDEs enjoy seemingly complete recall. Further, numerous studies have also shown that their recollections of what happened during their resuscitation are incredibly accurate. Conversely, when patients with no claimed out-of-body experience are asked to describe their resuscitation—even if that means guessing—they get it wrong. Television hospital dramas, it seems, are no substitute for *being there.* But there is another subtler yet important point that I think gets lost in the debate.

The real mystery is what happens to human consciousness after death—our memories, perceptions, and sense of self. Skeptics who adhere to a physical or materialist worldview argue that consciousness is purely a product of the brain. When the brain in my skull dies, so do "I." But the NDE has ultimately forced skeptics into adopting, unwittingly, the same position as believers: the NDE occurs independently of the brain.

Skeptics, more than forty years after Kübler-Ross declined to publish her own accounting of the NDE, are essentially arguing that the same experience can be had both with and without any number of drugs; in or out of the death state; and with or without a lack of oxygen—the same experience irrespective of whatever circumstances the brain might find itself in. This makes no sense from a materialist perspective. Drugs, anesthesia, a lack of oxygen, the flood of chemicals released by the brain after the heart stops beating—all these factors have known effects on the nature and quality of brain function. Skeptics, no doubt aware of this, often contend that many of the experiences lumped together under the umbrella term "NDE" are somehow different from one another. But thus far, they have failed to produce data that comprehensively demonstrate how the NDE takes on predictable, quantifiable changes in content and character, concurrent with the state of the body and brain.

Materialists take heart. This doesn't mean there is an afterlife. But it does mean the source of the NDE remains a mystery and could yet be proven an illusion or reality. Something is at work here, but what? I argue that, at the moment, we don't know. We understand far too little about the experience, and the reports of experiencers yield far too little actionable data for anyone to firmly conclude the NDE represents a real glimpse of the afterlife. But as yet we have no firm materialist theory to explain the NDE away.

So where does this leave us?

Well, as noted metaphysician and philosopher Terence McKenna put it, in a *Wired* interview conducted shortly before his own passing, all argument about our mortality comes to naught. Death remains the vast black hole of biology: "Once you go over that event horizon, no messages can be passed back. It represents a limit case in the thermodynamics of information. So what is it?"

It hurts, I think, for all of us—believers and skeptics alike—to admit it. But in answer to McKenna's question, all we can say is, *We don't know*. The black hole is just . . . black. We can, however, learn something about ourselves by taking a look at how we react to all these arguments. And we can best begin to glimpse this insight, I think, by looking at a powerful modern example of both skeptics and believers seeing only what they want and expect to see.

The occasion was a 2004 study conducted by Willoughby Britton, then working on a doctoral thesis in psychology at the University of Arizona. Britton had been reading about theoretical connections between epilepsy and paranormal experience. Some epileptics report they hear heavenly music or have religious visions just prior to seizure (though this happens far more rarely than skeptics would have us believe). The seizures themselves are triggered by mass firings of neurons in the temporal lobe, so Britton thought she might look for a connection there with NDEs.

She knew this sort of research was fraught with implications for her career, all of them bad. "The paranormal isn't supposed to be discussed," she says now. "It isn't supposed to be studied."

She brought the idea to her supervisor, thinking he would shoot her down. But to her surprise, he agreed. "I've made my reputation," he told her. "What are they going to do to me?"

Britton's idea was to find people who had experienced NDEs and monitor their brain activity overnight. In the end, she found just twenty-three subjects—not, by her own estimation, a large enough sample size to render her study authoritative but enough to start developing a picture. She also enlisted twenty people who had *not* undergone NDEs, or any life-threatening events, as controls. She believed that people who had undergone NDEs might show the same altered brain firing patterns as people with temporal lobe epilepsy. And five of her subjects did show these altered firing patterns. None of them were in fact epileptic, but they did display sudden spikes of neuronal activity. In the average population, according to Britton, the researchers might well have turned up no one with such abnormal brain activity, but among those who experienced NDEs they had found five—or 22 percent.

That meant her findings were, from a statistical perspective, highly significant. Still, she also knew her study's limitations: a small sample size, conducted over just one night, with no baseline reading for the people who had undergone NDEs. Without knowing what their brain activity looked like before the experiences, it's impossible to know what was cause and what was effect. "I thought the study suggested it would be worthwhile to conduct further research," she says today. "I thought it suggested there might be some link between the temporal lobe and the experience. But I also thought it was a pretty humble study."

She was shocked at the attention she received, which included articles in the *New York Times* and *Discover*. She also didn't expect that people would interpret her work to be so authoritative, precisely because she knew it wasn't. "I guess, usually, you'll find scientists arguing for the importance of their research," she says. "But people really went overboard with this."

They went overboard in arguing for her study's significance, and worse, seemed to interpret her findings in a peculiarly biased fashion. Britton thought both the *Discover* and *Times* articles "subjected the people in my study to being told they were dysfunctional."

This was deeply ironic. Because in the psychological questionnaire Britton gave all her participants, those who experienced NDEs rated more highly than the controls on measures of "active coping." What this means is that people who experienced NDEs were more likely to handle life's problems directly—without waiting for time, someone else, or a miracle to do it for them. They were, in comparison to the control subjects who had no NDEs, more optimistic and aggressive in their approach to building the lives they wanted. From my perspective, however, as a reporter, I can understand why these stories turned out as they did. The default position for media covering the paranormal is gentle, learned skepticism. In these media formulations, the NDE isn't a profound life-changing experience but an abnormal firing of neurons. And people who experience NDEs are to be characterized in kind.

But what most stunned Britton were the letters she received afterward. "I got letters from people who told me, 'Thank you so much for proving the Near Death Experience is real,'" she says, even though she felt she had done no such thing. "And I also got letters from people, including colleagues, *scientists*, saying 'Thank you for proving there is no afterlife and religious belief is a brain disorder.'"

Of course, she hadn't studied religious belief at all. But to some people all paranormal claims are supernatural claims—and both belong in the same dust basket. So in this instance, even scientists weren't immune to seeing things in Britton's study that weren't there.

We could choose to see the fallout from Britton's research as evidence of all that divides us. Some of us believe in the paranormal and see it in every coincidence. Some of us believe in a strictly materialist interpretation of the universe and see believers in the paranormal as retreating into comforting superstitions. But I believe Britton's study didn't reveal a fissure in our society—or at least, she didn't *just* reveal a fissure. I think what she revealed is a stunning point of relatedness.

Believers and skeptics alike tend to look at the paranormal and see what they want to see; they look at the paranormal and see a reflection of their own worldviews. And so what she calls her "humble study" and her "agnostic data" becomes confirmation in the eye of a passionate beholder. *Whatever* it is they're passionate about.

We should hover over this fact for a good, long while. We are, each and every one of us, just trying to get from day to day as best we can, with as little pain as possible. In this deep and abiding similarity, we each fight to maintain our worldviews. And we become irrational in various areas of life, from the paranormal to politics. In the year after President Barack Obama's election, in fact, polls were conducted that showed both Democrats and Republicans held withering opinions of each other's presidents. Big chunks of the Democratic electorate believed George W. Bush had some foreknowledge of the attacks of September 11, 2001, and allowed them to happen. An even higher percentage of Republicans believed Obama to be a socialist, questioned his citizenship, and wanted to see him impeached. Nevermind the facts. When someone advances an idea inconsistent with our own worldview, we don't just disagree—we start painting a mental picture of the person we oppose as somehow deficient, all higgledy-piggledy in the temporal lobes, perhaps, or just an outright villain.

And the beat goes on.

Just as I was finishing this manuscript, a study was released in which a small correlation was found between elevated carbon dioxide levels and the occurrence of NDEs. The study sparked headlines of the "NDEs Explained" variety. But anyone who has reviewed the literature can immediately recognize this research as a total outlier. Number one, the sample size was just eleven people, and their CO_2 levels were only slightly elevated—in fact, scuba divers can have similar carbon dioxide levels, but they don't go around claiming they swam their way to God. More important, numerous studies had already found that there is no apparent link between an NDE and heightened levels of carbon dioxide.

Jeff Wise, a science reporter for *Psychology Today*, was among those who wrote enthusiastically about this new carbon dioxide study, claiming it as an explanation for the NDE. Alex Tsakiris invited him on his podcast, *Skeptiko*, to discuss why he granted so much authority to research with so little actual weight. For the most part, the discussion was completely cordial. But toward the end, after Tsakiris had gently schooled Wise for about twenty minutes, the reporter broke in with a question of his own. "Are you," he asked Tsakiris, "a creationist?"

His implication was clear: anyone arguing that the NDE remains unexplained must not believe in evolution, must be anti-science. And that pretty much captures the tenor of the debate between believers and skeptics, each side harboring ill opinions of the other, each side making strange assumptions about the other's beliefs.

Elisabeth Kübler-Ross made history by initially avoiding such debates altogether. But before long, she would be drawn into them. And the personal consequences she suffered were tremendous.

IIIIIIIIIII

WHEN *ON DEATH AND DYING* was published, the changes in Elisabeth Kübler-Ross's life came quickly. She had written an unlikely bestseller, a book that became, seemingly overnight, part of the canon of course work for medical and nursing students; and, she reached the people who watched loved ones die in a soulless, dehumanizing fashion. Everyone wanted a piece of her. And in time, she visited nearly every continent and received twenty honorary degrees.

Her son Ken remembers the biggest change came in the form of a giant U.S. Postal Service sack that arrived every few days, bearing hundreds of letters. "My mother thought it was important to respond to every letter, personally," he said. "I didn't feel abandoned. She took me on trips with her when I was out of school. But it *was* different. She tried to do all the things we did before, and called us on the phone every night when she was gone."

Speaking to Ken Ross, I can't help but notice that his normally ebullient tone drops an octave when he talks about those days. Kübler-Ross's sudden, international celebrity created a massive strain on her marriage. And she had also developed a kind of paranormal problem. She didn't publish her own experiences with NDEs in *On Death and Dying*. But she kept working at the bedsides of the sick, and strange events did not seem to leave her be. Less than two years later, in fact, she had her strangest experience of all, which she documents in *Wheel of Life*.

The constant lectures, seminars, and out of town speaking engagements, the hours spent attending to the dying, had worn her down. She was considering giving up the work.

She stood in the Chicago hospital where she became famous, talking to a colleague, when she noticed a woman near the bank of elevators. Kübler-Ross had been deep in thought. But this woman caught her attention. She thought she had seen her, somewhere, before. Then she noticed something alarming.

The woman was semi-transparent.

After Kübler-Ross ended the conversation with her colleague, the woman approached her, not walking so much as floating. "Do you mind if we walk to your office?" the apparition asked.

Kübler-Ross said yes, and started the strangest walk of her life, a few dozen yards to her office. Inside, she remembered the woman's face. It was Mrs. Schwartz. She sat down, thinking she might faint. Schwartz had died some ten months earlier.

Kübler-Ross questioned her own sanity.

In the course of her work she had counseled schizophrenics. And when they saw something that wasn't there, she didn't feed their fantasies. She told them they were hallucinating. So she reached for her pen, her papers, her coffee cup. She tried to tether herself, through touch, to the real world. But the apparition didn't fade at the great lady's attempts to make her go away. In fact, the spook spoke. "I had to come back," Schwartz told her, "for two reasons. Number one is to thank you and the Reverend [Imara] for all you have done for me. However the second reason I came back is to tell you not to give up your work on death and dying . . . not yet."

Kübler-Ross wondered how Schwartz could possibly know she was planning on quitting. But she also continued to question whether the entire event was transpiring at all. At the ghost's behest, she promised not to quit her work yet. And in return, she asked the ghost for a favor. "Will you," she asked, "write a brief note for Reverend Imara?"

Mrs. Schwartz complied, taking a pen in her hand, then disappeared. We'll get back to that note. For now, understand her old patient's alleged reappearance as a ghost would become a prominent feature in the story of Elisabeth Kübler-Ross. But for the moment, it was private. Kübler-Ross only outed herself as an experiencer of anything odd at all in 1975 when the author Raymond Moody asked her to write the foreword to *Life After Life*—

the book that coined the phrase "near-death experience." She and Imara looked over his manuscript. The experiences their own patients recounted were accurately mirrored in Moody's own research. They also believed he had written credibly. But there was something more: Kübler-Ross was changing. "At this point in my life," she later wrote, "I was open to anything and everything. Most days I felt as if a curtain was being lifted to give me access to a world no one had ever seen before."

This was also the problem. Because just then, vulnerable in the confines of a strained marriage, a grueling career path that was wearing her down, and new scrutiny related to her own investigation of the paranormal, she received a phone call. The people on the other end were Jay and Martha Barham, who had been drawn to her by coverage they had seen of her controversial endorsement of Moody's book. They called her from San Diego and promised her something more incredible than she had ever experienced, something that, deep inside, she longed for. They promised to introduce her to spiritual entities. They promised to reproduce the strange, sporadic mystical experiences Kübler-Ross had enjoyed—and to do so on demand.

The Barhams had found a fertile target. And the esteemed psychologist quickly booked a speaking engagement in San Diego so she could take a side trip to meet the Barhams, who met her at the airport and hugged her like old friends. From there, they whisked her to the Church of Divinity, where Barham channeled spirits for a congregation of about a hundred people. "On my first day there," she later wrote, "I joined 25 people of all ages and types in the dark room [of a windowless building]. Everyone sat on folding chairs. Jay placed me in the front row, a spot of honor. Then the lights were switched off and the group began singing a soft, rhythmic hum that built to a loud group chant, which gave Jay the energy needed to channel the entities. . . . As the chanting reached a new, almost euphoric level, Jay *disappeared behind a screen* [my emphasis]. Suddenly, an enormously tall figure appeared to the right of me. . . ."

Over the ensuing months, Barham introduced her to spirit guides named Salem, Pedro, and an entity named Willie. This was strictly old-school, séance-style mysticism. In the 1930s people advertising themselves as psychic mediums turned out the lights and channeled the dead, asking the spirits to knock once for yes, twice for no. But really it was an assistant in on the gag or the medium making all the noise. Like the mediums of old, Barham, too, insisted all the lights be turned off—lest he or the spirits be damaged by the terrible power of the 60-watt bulb.

Kübler-Ross should have known better, but she was too vulnerable, it seems, to see straight.

Attempts were made to save her.

Her husband Manny answered the phone once to find what he took to be a man disguising his voice on the line. The man claimed to be Kübler-Ross's spirit guide. Manny hung up. How, he asked his wife, could she possibly fall for such obvious bullshit? But fall for it she did. She had quickly grown dependent on Barham and his church. And when she wouldn't give up this newfound mysticism, or the constant lectures, Manny asked for a divorce.

Ken says it was clear "they never stopped loving each other." And because of his mother's hectic travel schedule, the kids lived mostly with their father. Without her husband or children, Kübler-Ross fled to the Church of Divinity. She acquired a parcel of land nearby and started a center she called Shanti-Nilaya, meaning the "final home of peace." She put Barham to work as a full-time spirit channeler.

Manny still worried about her and made a phone call of his own—to Imara, who traveled to the center to investigate. He found the landscape around the center incredibly beautiful—forty acres of swaying trees and lakefront views. But what was happening inside was "pure evil."

On the very first night, he sat through a séance. His friend had fallen for what he calls a "bad acting job." But he was chilled, the next day, when he walked through a hallway in the center and noticed Barham in a common room, staring intently at a television with the sound turned down. Imara stood in the doorway and says he knew, *felt*, just what Barham was up to—immediately and in his bones. "That sick bastard was practicing reading lips," he says. "Who *ever* does that?"

Was this one of Barham's means of gaining supposedly "psychic information"? Imara didn't care. He just wanted to get his friend away from the man, and tried, several times, to share his own observations with Kübler-Ross. But every time he said something negative, she interrupted him. She changed the subject. This told him all he needed to know. The Elisabeth Kübler-Ross with whom he worked had never interrupted people.

She listened.

In old videos of her at the bedsides of the dying, Kübler-Ross can still be seen gazing so intently at the sick as they speak that she seems to have no other possible purpose but to serve as a kind of universal mother—the receptacle for her patient's woes. But this woman seemed trapped inside herself. "I think it was all internal," Imara says. "This was about where she was and what she needed at the time."

He also shared with me an observation that is key to understanding the paranormal—and coming to grips with the paradoxical tale of Kübler-Ross, the insightful psychologist who seemingly lost her own grip on reality. "The things we saw," he says, "had been incredible, but they were sudden, and you couldn't count on them."

The paranormal, as we'll find throughout this book, simply doesn't ever occur on demand. We move through this life on life's own prosaic terms, mostly. In the materialist formulation, we feed our stomach-furnaces with food to sustain us. We input and output information using our computer brains. We live in a Newtonian realm, where most everything, most all of the time, moves in precise, predictable, patterns. Kübler-Ross had, like many others before her and since, come to need someone or something in her life that could recreate those unexpected glimpses of something *other*—that could, metaphorically anyway, show her the tunnel of light she'd heard of in her patients' near-death experiences. And she needed it so badly, she could plainly no longer see past her hopes and wishes—to spot even the most obvious con. And so she looms, I think, as the ultimate cautionary tale for all those who wish to explore the paranormal: Here be dragons. And by 1979, they were at her door.

Time, *People*, the *Los Angeles Times,* and *Harper's* magazine all wrote in-depth accounts of Kübler-Ross's fall. And for a while, Kübler-Ross defended herself and Barham. She began to share the Mrs. Schwartz story at lectures and talks and endorsed all manner of nonsense. Things were bad, and they got worse. "Appearing at sessions in darkened rooms . . . ," reads *People*, "these 'entities' [Barham channels] have assumed human form and according to some reports engaged in sexual relations with church members."

Church congregants noticed things they should have seen early on, things that had been painfully obvious to Imara. The supposed entities smelled of cigarettes, like Barham, and spoke with similar accents, pronouncing "escape" as "ex-cape," for instance. Five women, each of whom thought they had sex with an afterlife entity, came down with the same vaginal infection. And finally, one of them did the unthinkable. She turned on the lights during one of Barham's channeling sessions. The entity that had supposedly materialized was Willie. But when the lights came on there was no spirit. There was just Barham. Naked except for a turban.

Kübler-Ross hung on to her fantasy for a while longer. But when Barham's explanations no longer satisfied her, she parted ways with him, the great psychologist now the victim of a very long con, carried out over four years.

A skeptical mind most certainly would have been helpful, a good, ruthless debunking was called for; because a too-open mind had ruined a reputation.

||||||||||||

KEN ROSS, THE GREAT lady's son, now looks after his mother's foundation. When someone like me comes along, looking for information, he serves as gatekeeper. HBO worked with him as they tried to develop a script. The latest to give it a go is actress Melina Kanakaredes, formerly of CSI: New York, who was developing a script with the cooperation of Kubler-Ross's estate as I put the finishing touches on this book. She faces a significant challenge. Because Kubler-Ross's life won't submit itself to a neat, one-hour-and-forty-minute retelling.

In her life's final acts, after Barham, she seemingly reinvented herself yet again, working with AIDS patients throughout the 1980s, the new disregarded among the terminally ill, who faced a level of stigmatization perhaps not seen since the lepers of biblical times. Then began the series of debilitating strokes that ultimately ended her life. The story is difficult to tell not only because of all the events packed into it, however, but because the twists and turns could polarize an audience.

What do we do with the paranormal stories that accumulated around her? What do we do with . . . *the note?*

Remember, Kübler-Ross asked the ghost of Mrs. Schwartz to write a note to Imara, the story goes, and the semi-transparent being complied. According to Imara, Kübler-Ross did hand over the note—it was addressed to him, after all—and he owns it still.

People have asked him for copies over the years, intending to analyze the handwriting, to compare it to a sample from Schwartz. An attorney from Florida recently implored him to submit the note for investigation. But Imara said no. "One more argument behind that woman's name is not something I'm going to contribute to," he says.

His math is simple, and unassailable, and provides a fitting end to this tale: if the handwriting doesn't match (as seems to me by far most likely), skeptics will use that information to undermine the veracity of everything Kübler-Ross ever said. But as Imara puts it, the "story about Mrs. Schwartz is nothing. It is inconsequential, in comparison to the many things that happened. Things I witnessed myself."

Believers, in turn, would argue, less logically, that a change in Mrs. Schwartz's handwriting style might reflect some effect of, I dunno, crossing over? But, as unlikely as it seems, what if the note *did* match some existing sample of Schwartz's handwriting? Well, even that wouldn't get us any closer to real answers: skeptics would just holler fraud—or more profitably point out that handwriting analysis is downright subjective, arguably not science at all.

This tale ends, then, in the mire, the debate so polluted that sometimes, for the people involved, it seems best simply to permit no further inquiry. And so the life of Elisabeth Kübler-Ross will continue to be punctuated both by exclamation points and question marks—the sum of a life spent collecting the last stories of the dying.

2

Do You See What I See?

The Curious Conflict Between Telepathy Skeptics and Believers

Because something is happening here. But you don't know what it is. Do you, Mister Jones?

—Bob Dylan, "Ballad of a Thin Man"

In 2001, British physicist Brian Josephson was asked by the Royal Mail, Britain's postal service, to write a short essay commemorating a new series of Nobel Prize–themed stamps. He could have just written the standard thing—extolling the virtues of science and urging kids into the field. But what he delivered, and the Royal Mail published, deviated more than a few degrees from standard. "Quantum theory is now being fruitfully combined with theories of information and computation," Josephson writes. "*These developments may lead to an explanation of processes still not understood within conventional science such as telepathy, an area where Britain is at the forefront of research.*"

The mention of telepathy, invoking the paranormal, caused a furor. And in response, some of Josephson's fellow physicists railed to the press, accusing Josephson of having "hoodwinked" the Royal Mail into printing falsehoods.

Josephson, a Nobel Prize winner, is an avowed believer in telepathy. (In this chapter, I use the umbrella term *psi,* which includes telepathy and covers any theoretical ability to gather accurate information outside our five normal sensory pathways.) Claims of psi-ability have been with us for millennia. The Greek historian Herodotus reported that, in 550 B.C., the Oracle at Delphi predicted precisely when the king of Lydia would be boiling a lamb and a tortoise in a brass cauldron. Hardly as valuable as predicting, say, the winner of the Super Bowl. Still, the Oracle got gold and silver for her trouble.

Today, in the modern West, psychics can also earn their share of filthy lucre. But the mainstream view of psi is contentious, to say the least, and Josephson's full-bodied embrace of psi is surprising—not least because he could easily have continued down the less nettlesome path he had forged for himself. As a graduate student, Josephson had correctly predicted that a phenomenon called "quantum tunneling" was more powerful than previously thought. His research led to Josephson Junctions, in which two layers of superconducting material sandwich a (very) thin layer of nonconducting material. This construction allowed electron pairs to "tunnel" from one side to the other, leading to a vast array of practical applications, like microchips and MRI machines. In short, Josephson's discovery is among the most important technological leaps of the past half-century. But because of his interest in psi, some now portrayed him as a figure of disrepute. He had gone "off the rails," they claimed in the wake of his offensive sentence, his intellect somehow damaged by his long-running study of telepathy.

"It is utter rubbish," David Deutsch, a quantum physics expert at Oxford University, told the *Observer* newspaper. "Telepathy simply does not exist."

BBC Radio invited Josephson to defend his position against two skeptics—the key one, for our purposes, being American James Randi. A former stage magician, Randi has been debunking all things paranormal for roughly forty years. And given the opportunity to confront Josephson, he attacked. The magician accused the physicist of invoking the "refuge of scoundrels" in referring to quantum mechanics and further claimed there was "no firm evidence" for telepathy a reputable scientist would accept. But there is a problem here. Because the evidence submitted for psi is vast, and so competently assembled, some more fair-minded skeptics have been forced to concede important ground. "I agree that by the standards of any other area of science that remote viewing is proven," says psychologist and skeptic Richard Wiseman, in a January 2008 edition of the *Daily Mail*.

Remote Viewing (RV) is the claim of a real Mind's Eye—the ability to see things and describe them accurately without being bodily present to see them at all. This seemed a startling admission. And Wiseman was asked to clarify. As expected, he claimed to have been misquoted—but not in the way we might think. He wasn't referring *only* to remote viewing, he said. He was describing the entire field of telepathy—or psi research in general. What's more, fellow U.K. skeptic Chris French *agrees* with him. "I think Richard's right," he told me. "For an ordinary claim, the evidence we already have would be sufficient."

The issue, as described by both Wiseman and French, is that telepathy is no ordinary claim. The finding of an as yet undiscovered sensory capacity might force us to question all kinds of scientific truths—in physics and neuroscience, just for starters. So, the thinking goes, the evidence provided for telepathy must be as extraordinary as the claim itself. Which, they maintain, it isn't.

It seems, then, when the evidence for psi is closely reviewed, the reputation of Brian Josephson can safely be removed from hell (the Landing Place of Scoundrels) and cast into purgatory (the Landing Place of Stuff We're Still Debating). What we'll learn here, in the muddled middle, is that psi proponents and naysayers seem diametrically opposed—like rival families in the Ozarks, pop-eyed with adrenaline that can only come from really, really wanting to shoot someone else in the heart. But truth be told, the skeptics are far more like the believers than they first appear. And in a very real sense, psychic slayers and psychic supporters are, shockingly, both right.

IIIIIIIIIII

I WALKED TO MY first session of the Parapsychological Association's (PA) summer 2009 meeting through a light Seattle mist. Though I knew that most of the leading researchers in the field of parapsychology would be here, I didn't know quite what to expect. This was an academic conference, and the talks promised to be incredibly technical. But I did have an impromptu introduction to one of the weekend's speakers. I had checked into a dormitory on the University of Seattle campus the previous night and run into a big, bald-headed man out by the elevators. The hallway was cramped, and we were ostensibly here for the same reason, so I introduced myself.

"How you doing Steve!" the man replied, pumping my outstretched hand as if we were old friends. Before I knew it, I'd been invited to sit at a desk in the tiny dorm room of Paul H. Smith, a former military man who had carried out remote viewing trials for the U.S. Army.

I'd come here, I thought, with an open mind. I'd read the scientific literature on remote viewing, pro and con. I knew the Army really did have a remote viewing program, investing money and facilities in a network of psychic spies. But until I sat with Smith, those were just words on a page. Remote viewing is such a strange idea. Theoretically, a viewer can "see" things with his or her mind—no matter how many miles away, underwater, inside cabinet drawers, and even in outer space. Not even time is a factor, as proponents of remote viewing claim to see the past and the future. Confronted by Smith, I was a little dumbfounded that this big, matter-of-fact Texan was here, talking to me about mind sight.

"I was skeptical about it, too, when I started," Smith told me. "But I wanted to see if it was real."

I wondered if I was smirking at him. Because seeing things at a distance just didn't fit into my own personal experience, and seeing someone *up close* who claims to see things at a distance felt even stranger. I'd imagined that recruitment into this odd military experiment was conducted with equal weirdness. Perhaps the sergeant in charge read his morning coffee grounds? But Smith had been brought aboard after a couple of innocuous conversations with a neighbor on the Fort George G. Meade military base. The neighbor was involved in the remote viewing program, a fact he concealed from Smith. But when he saw Smith had some talent for drawing, he tapped him for a tryout. Over the next couple of years, Smith put his artistic talent to use—trying to see unknown targets with his mind's eye and sketch them for his superiors. He had his share of success, he told me. And on that first night in Seattle, we talked for more than an hour—me mostly feeling odd that Smith himself seemed so credible, a plainspoken Texan who retained the upright bearing and polite demeanor of a career soldier.

The next morning, after worrying about whether or not my skepticism had shown, I kept looking in vain for some crowd of people all headed in the same direction, figuring they would lead me to the first conference session. But no crowd ever materialized. And it was only after I saw Smith walking across campus with the steady gait of a soldier that I found my way. The conference organizers at the Parapsychological Association had warned me the gathering would be small. Just how small came as something of a shock. I counted maybe thirty people on hand for the morning's first presentation, a panel dealing with the role that belief systems play in science. "Dick Shoup continues to work on psi-related issues, undistracted by any significant funding," read the bio of one of the panel's participants.

The line got a nice laugh from the audience and set the tone for a real budget catering affair, all of it held in one college lecture hall, with a table out front at the breaks holding bottles of water, raisins, and the occasional cookie. In my capacity as a journalist, I've attended enough professional functions to know where the money is (trial attorneys throw the fanciest parties). And clearly, I found, there is no money in psi research.

On one level, this might not be so surprising. After all, the participants had arrived to spend the weekend discussing a phenomenon many don't believe exists. Perhaps more commonly known under the heading of ESP, or extrasensory perception, the Greek letter *psi* is employed by physicists to depict a quantum mechanical wave function, and by paranormalists to cover four main areas of research: telepathy, or a connection between separate minds; clairvoyance, or "clear seeing," which would encompass remote viewing; precognition, somehow perceiving events in the future; and psychokinesis, or PK, which might mean bending a spoon or influencing the output of a random number generator, affecting the physical world using only the power of our consciousness.

The attempt by paranormalists to co-opt a term employed by physicists probably speaks to two things: one, they think the mechanism by which a "sixth sense" works is physical, and scientific, not immaterial and unverifiable; two, they need, want, wish, *long for* the imprimatur of mainstream science.

In his 1979 paper, "Experimental Parapsychology as a Rejected Science," University of Pennsylvania sociologist Paul D. Allison surveyed members of the Parapsychological Association. What he found was a small group of dedicated and demonstrably qualified psi researchers who reported that they had been routinely discriminated against by their mainstream, university bosses in hiring, promotions, publications, and funding research.

Thirty years later, after eating my share of raisins at the PA's small, dispirited annual conference, I wondered if anything had substantively changed. I contacted Allison, who dug up a draft of the questionnaire he submitted. I reshaped it as an online survey, had a techie friend slap it on the Internet, and asked the PA's current membership to respond.

What I found is that the relationship between psi researchers and mainstream science remains the same: more than half of my subjects felt they had been discriminated against or were aware of some kind of discrimination having been waged against a psi researcher. That figure was down only marginally from Allison's survey. Further, nearly half of my respondents felt they had been denied funding or facilities for the crime of studying psi. I received just thirty responses, less than half of what Allison got. But these self-reports matched the downbeat parapsychogical procession I had witnessed with my own eyes. And maybe that *is* to be expected. Maybe parapsychologists don't *deserve* funding. Maybe this state of affairs shows how well science works, and nothing has changed substantially in the way parapsychologists are treated because their findings remain stagnant. But that seems an unlikely answer. Because their findings have changed—a lot.

There are several major areas of psi research. But I'll quickly sketch out three here. (Readers interested in exploring the research personally should consult my Notes and Sources section, which includes information on many studies, the vast majority of which were published in the past twenty years.) Let's start with remote viewing, one of the most prominent and successful areas of psi research to emerge in the past three decades. The military conducted a lot of its own experiments to see whether remote viewing worked. The protocol the Army employed went something like this: a viewer in training is set up at a table. His supervisor gives him some minimal piece of information, like latitude and longitude coordinates, or merely tells him to focus on "the target," without being given the foggiest clue what the target might be. The viewer is supplied with a pen and paper to record his or her presumably psychic impressions, usually in a sketch. In these trials, any written impressions produced by the viewer—in words or pictures—are given to an independent judge. The judge is also supplied with four photos, one of which displays the target. The judge, blinded to the true target, then matches the viewer's report to the image he feels it most closely resembles, giving any viewer a 25 percent shot at scoring a hit purely by chance.

In practical applications of remote viewing, the Army simply took what information the RVer provided and looked for anything that might contain actionable intelligence. Few dispute that the Army's remote viewers occasionally scored remarkable hits or dazzle shots along with their misses. Pat Price accurately described a military installation, including some current and past activities and even the site's code name. Joe McMoneagle, one of the most (in)famous RVers, was asked to see the content of an airplane hangar. In this instance, he was even given a photo of the hangar's exterior. His handlers were trying to fool him, figuring he might start drawing pictures of airplanes. Instead, he drew a tank parked inside the hangar, accurately depicting the vehicle's interior, including laser range-finding equipment, visual systems to compensate for low-visibility conditions, and cutting-edge computers.

McMoneagle also accurately described the contents of a building on a Soviet naval base on the Baltic Sea. Military analysts initially scoffed at what he came up with—a submarine far bigger than any then known, with a set of missile tubes located, contrary to standard design, in front of the conning tower. But later, satellite photos proved him right. He had apparently described the *Typhoon*, a super-secret Soviet sub.

Skeptics might explain these kinds of dramatic successes as the product of basic probability laws—make enough drawings of enough targets and you're bound to get something right. Or perhaps the details, in these particularly evocative cases, had somehow been leaked to the viewers. But this is far from the last word. The Army's RVers didn't just capture dazzle shots. In fact, analysis of their work suggested they were producing accurate information at a rate significantly above chance. And in one analysis, skeptics and believers came awfully close to lying down together. In fact, skeptic Ray Hyman wrote in his "Evaluation of a Program on Anomalous Mental Phenomena" that the results he saw could not, in his estimation, be put down to chance or any methodological flaw he could find. He refrained from calling psi an established fact by only the thinnest of margins—positing that some methodological flaw might be discovered in the future.

Since then, the state of play has remained much the same, and the storm of debate in psi research has often revolved around meta-analyses. A meta-analysis is, in essence, a study of studies. Take a stack of related research findings, conduct a rigorous statistical analysis, and read the numbers. Well, meta-analyses of what's known as the Ganzfeld database tend to show evidence for the unbelievable. In Ganzfeld tests, receivers sit with halved Ping-Pong balls taped over their eyes and plugs in their ears, then try to pick up accurate information. Proponents say that cutting off the normal sensory channels allows other information to come through—like a radio antennae plucking the signal from noise.

Performing a meta-analysis on this research, one of the most well-known scientists working in parapsychology, Dean Radin, found a 32 percent hit rate when 25 percent is expected by chance. He calculated the odds against chance for the positive results at 29 quintillion to 1. But the single most promising area for further research may lie in "brain correlation" experiments.

These studies go by different names, which rigorously avoid the dreaded *T* word, because the mere mention of telepathy is likely to draw emotional fire from someone's amygdala, somewhere. (*Case in point:* British parapsychologist Guy Lyon Playfair tells a very funny story about attempting to study psychic functioning in twins. Instead of using the word *telepathy*, he tried to placate the mainstream by declaring he was studying "biological correlates of empathy.")

In most of these studies, two subjects, usually with some prior personal connection, are separated and rigged up in skullcaps designed to monitor their brain waves. While one subject, the "receiver," sits in a bland featureless room with nothing much happening, the "sender" is exposed to stimuli.

The idea is to see whether the introduction of a stimulus to the "sender" produces a corresponding reaction in the receiver's brain. Will agitation in the sender, or for that matter pleasure, be mimicked in the brain of their partner? Dean Radin has collected thirty-three of these experiments, the majority conducted since 1994, with strongly positive results for the presence of brain correlation—which we'll bravely call telepathy.

This is a drop in the scientific bucket, of course. When a field is really accepted, studies might more likely number in the hundreds—another sign of how psi remains ghettoized, or how far it still has to go.

There are small signs we might someday see a shift. Earlier, I mentioned that skeptics Wiseman and French had been gracious enough to admit the obvious: psi researchers have lots of good evidence. The question is whether or not they have enough. Perhaps even more incredibly, one of this country's most vocal atheists, Sam Harris, in the *End of Faith*, acknowledges the research for psi is so compelling that telepathy may need to be admitted into the canon of accepted knowledge. And there is more. Along the way, I spoke to Dr. Michael Persinger, a real gadfly to paranormal believers. Persinger is in most respects a kind of happy naysayer—no to God, no to ghosts—but he also happens to be a strong proponent of psi; and he was in the process of publishing his research on what he calls "the Harribance Effect."

Working with an under-the-radar psychic named Sean Harribance, Persinger claims to have found a pattern of brain activity that correlates with psychic functioning. "Here's the really exciting part," he says. "Here's the wow. When Harribance has actually gotten correct information, his brain state corresponds demonstrably with that of the person he's reading."

Harribance spoke to me at length over the phone but wouldn't agree to see me in person. "I'm not interested in publicity," he told me, multiple times.

Clearly, something is happening here in the land of psi. And inexplicable results should be the ones that pique a scientist's interest most. But once a subject has acquired what I call "the Paranormal Taint," mainstream science tends to run in the other direction. As I see it, the Paranormal Taint is itself harmless, inert, and value neutral. But some people still regard the Taint either as a sign of holiness or toxicity, depending upon their point of view.

Both camps—of passionate believers and equally passionate deniers—are surpassingly small. Most of us are likely in the middle. We don't think much about a subject like psi research from day to day, and whatever opinion we hold is provisional. In other words, we would be interested in hearing out the evidence. The problem we encounter is the same one we see in our politics: When the media investigates a phenomenon like psi, or for that matter privatizing Social Security or forming a public health care system, they reach out to sources with diametrically opposed positions. That makes for higher drama, more colorful quotes, and, so the thinking goes, better radio or television.

What it doesn't bring us any closer to is the truth.

In Seattle, I met a woman who has been looking at the totality of psi research as dispassionately as she can, under the circumstances. Jessica Utts is a statistician from the University of California who admits she has always had an open mind about psi research. Her father had conducted some informal studies with her when she was a kid, she told me, with some success. But later on, when she delved into the world of statistics, she realized just how little such a home experiment mattered. "It wasn't science," she told me, "and of course there was too small a sample size. We might have just gotten lucky. But it stayed with me, as a matter of curiosity."

Utts has since gone on to write mainstream textbooks in her field, including *Seeing Through Statistics*, a primer on the responsible and irresponsible uses of statistics. She is aware that numbers can be abused just like words can—to say anything we'd like. And she learned to be a skeptic.

We spoke in a large cafeteria on the university campus. Students taking summer classes sat around us, out of earshot. Utts, a matronly looking woman with big eyes, a kind demeanor and a bob haircut, sipped a soda and told me about her first public foray into the paranormal.

Someone had advanced the claim that fewer riders board trains that subsequently crash. He had conducted a statistical analysis he felt demonstrated this fact, and a TV station investigating his study called Utts to look over his handiwork. In short order, she found a crash crucial to his finding occurred on Labor Day, when the vast majority of people were off work. Once the Labor Day crash was factored out of his study, his evidence evaporated.

Eventually, Utts applied her statistical expertise to the field of psi research. She found mixed results, which nonetheless topple the current mainstream view.

According to Utts, research claiming people can influence the output of a random number generator isn't convincing. Much of this research occurred at the Princeton Engineering Anomalies Research lab, or PEAR, and Utts thinks these supposed mind–machine correlations don't prove the ability of mind to affect matter. But remote viewing? Telepathy? She felt the body of research there was so compelling, both methodologically and statistically, that someone needed to speak up for the bedraggled community of psi researchers. And that's how a statistician came to be one of the leading members of the Parapsychological Association.

She knew involving herself could cause her trouble at work—and she wasn't wrong. In fact, she told me, whenever her contract comes up for renewal she faces pretty much the same scenario: the panel that decides on merit increases splits its vote because, at any given time, two or three people in her department find their skin itches when they get a load of her Paranormal Taint. Her fate hanging in the balance, her name gets kicked upstairs for a review, and . . . she gets paid. My own estimation: her mainstream credentials—those textbooks—are so strong that her credibility can't be denied. "Statistics," says Utts, "when you know how to read them and you have a big enough database, don't lie. I'm as convinced as I need to be that something is going on here. How it works, I don't know. You'd have to talk to someone else about that. But if you look at the research, the numbers are there."

||||||||||||

THE PHRASE "EXTRAORDINARY CLAIMS require extraordinary evidence" was first coined by skeptic Marcello Truzzi, of whom we'll hear more later. It was subsequently popularized by Carl Sagan. Regardless, the slogan is merely a pithy, modernized exaggeration of what the humanist philosopher David Hume declared roughly 250 years earlier about miracles. Hume famously argued that before we believe in a miracle, there should be so much evidence it occurred we'd be more foolish *not* to believe in it.

In practical terms, this makes good sense. Any claim that might require scientists to double back and reverify earlier findings or assumptions is a potential time waster. Before they begin questioning the foundation on which current research stands, they should see evidence that the ground really has shifted under their feet. The argument put forth by skeptics is that psi is an extraordinary claim; thus an extraordinary amount of data is required to support it. As skeptic Ray Hyman put it, if psi exists, "the fundamental principles that have so successfully guided the progress of science from the days of Galileo and Newton to the present time must be drastically revised."

For the moment, without accepting or rejecting Hyman's claim that psi undermines our understanding of physics, let's just understand that this is the calling to which skeptics claim to respond. And so they have insisted, for decades, that parapsychologists must employ tighter and tighter controls on their studies—to eliminate obvious possibilities like fraud, and more subtle ones, like sensory leakage, in which the receiver in a telepathy study becomes aware of the target. In response, parapsychologists have increased the rigor of their methodology. But no matter what lengths they have gone to in order to satisfy the skeptics, the skeptics have yet to be satisfied. And during a period in the late 1970s and early '80s, this back-and-forth between the skeptical community and parapsychologists was so robust that it became a subject of study all its own.

Using the conflict between skeptics and parapsychologists as a lens, sociologists began research into how science is conducted, not in its idealized form, but in reality. Sociologist Trevor Pinch subsequently identified a group of "scientific vigilantes," people who did not always hold scientific degrees but nonetheless appointed themselves to guard the borders of "true" science.

The sociologists involved never took a side on the issue of psi itself. But in mediating the debate, they described the skeptics as an unruly and largely unscientific bunch. I recently interviewed Pinch, one of the most prolific authors on the subject, and he told me his findings surprised him. Initially he suspected the accusations leveled by skeptics were correct: parapsychologists, as they were known, were somehow self-deceived, employing shoddy controls on their experiments or committing outright fraud.

What he found was the exact opposite: psi researchers took the skeptics seriously, conducting experiments according to methodology that at least kept pace with the most rigorous of the psychological sciences. When they produced positive results, the skeptics claimed the controls needed to be tighter still. Then tighter. "It was hard not to feel bad for the parapsychologists, really," says Pinch now. "These were qualified, sincere researchers doing serious work, and they always had to deal with this group of people that were essentially engaged in a lot of name-calling."

Clinical psychologist Elizabeth Mayer, in her book *Extraordinary Knowing*, tells her own story of discovery. She started looking at parapsychological research after a psychic correctly predicted where her daughter's lost harp could be found. She didn't believe in psychics. She had in fact learned during her own university education that parapsychology was bunk. And so she set out on a personal mission: to prove to herself that psi is hogwash. "I began discovering mountains of research and a vast relevant literature I hadn't known existed," she writes. "As astonished as I was by the sheer quantity, I was equally astonished by the high caliber. Much of the research not only met but far exceeded ordinary standards of rigorous mainstream science."

By way of comparison, she found the *Skeptical Inquirer*, one of the foremost skeptical journals, to read like a "fundamentalist religious tract." Like Pinch, she had quickly reached a surprising conclusion: professional skeptics accused the parapsychologists of practicing "pseudoscience," essentially a kind of fraud, in which psi researchers claim to be scientific but don't employ the scientific method. They just didn't seem able to support the charge.

The pseudoscience claim is often made of psi. But is there any data to support it? Well, in 2003, French sociologist M. C. Mousseau compared ten markers of good science with the work of parapsychologists. Her aim was to see whether psi researchers really are practicing pseudoscience.

She defined a *real* scientist as one who gathers or uses quantitative data, seeks empirical confirmation or disconfirmation, looks for correlations, relies on logic, proposes and tries new hypotheses, admits gaps in the current database, and is consistent with scientific work in other fields. She then reviewed the work in four "fringe" journals studying the paranormal and found them to be rigorously consistent in employing the methods of good science. In fact, when comparing the "fringe" journals to mainstream journals in physics, psychology, and optics, she found "no qualitative difference" between them. The fringe journals published fewer experiments, which she attributes to the relatively small number of practicing parapsychologists. (I saw just that, firsthand, in Seattle.) They do, however, publish a comparable amount of empirical data—and also question their own findings with an admirable openness. In fact, while parapsychologists regularly publish "null" results—studies in which no evidence of psi is found—the mainstream journals Mousseau studied published nothing but confirmatory data. In sum, then, when skeptics accuse parapsychologists of practicing pseudoscience, they are either lying; uneducated about what the parapsychologists are doing; uneducated about standard scientific practice; or, as I'm about to argue, engaged in a prolonged bout of self-deceit.

Chris Carter's *Parapsychology and the Skeptics* neatly captures the history of the schism between the two camps. That said, for the purposes of this book, the situation can be assessed with a few telling war stories, many of which include James Randi. Often called the godfather of skepticism, Randi has been a fount of both entertainment and valuable lessons in critical thinking. Think of him as a loud, small, gray-haired, and angry Velma from *Scooby-Doo*—always eager to pull the sheet off the supposed ghost and reveal the huckster within. He has successfully debunked psychic surgeons and faith healers. But far too often, he mixes his good work with bad, undermining the movement he purports to lead.

Does telepathy exist? Randi would say there is no valid evidence for it. But the truth is far more complicated. There is in fact good evidence to suggest psi is real. But as yet, there is no scientific consensus. What the ongoing furor over psi demonstrates, however, is that even rationalists can come to look like believers—motivated less by the data in front of them than by the worldviews closest to their hearts.

The foundation for the modern skeptical movement was laid by CSICOP, the Committee for the Scientific Investigation of Claims of the Paranormal. CSICOP was founded by humanist philosopher Paul Kurtz in 1976, when

Kurtz asked fellow skeptic Marcello Truzzi to join him as co-chairman. Truzzi published his own privately circulated newsletter at the time, *The Zetetic,* which served as a forum for skeptics and proponents of anomalous phenomena to engage in an ongoing dialogue. In contrast, Kurtz's own publication, *The Humanist,* seemed too shrill by half—tossing religious claims and well-controlled psi experiments into the same large dust bin. It is perhaps not surprising, then, that Truzzi was removed as editor of the group's new publication and left CSICOP shortly after joining it, convinced the board was more interested in promoting a polemical agenda than engaging in real inquiry.

Over the years, though he remained a skeptic, Truzzi thought the researchers turning up positive data for psi were on the whole ethical and practicing science. But "the problem with CSICOP is that it has made debunking more important than impartial inquiry," he later wrote. He even began using the term *pseudoskeptics* to describe the attitude of the worst offenders on CSICOP's board.

Things might have quieted down with Truzzi gone. But the internal purging grew bloodier. Dennis Rawlins, an astronomer enlisted into CSICOP, was the next to find his head on the chopping block. In a 1981 *Fate* magazine essay, Rawlins portrayed CSICOP as a gang of fanatics. And according to Rawlins's article, they began revealing themselves behind the scenes, in 1975, as CSICOP was forming.

The controversy started when prominent French psychologists Michel and Francoise Gauquelin presented evidence for what they dubbed the "Mars effect," which claimed a disproportionate number of European sports champions had been born with Mars rising in their astrological signs. Kurtz published a pair of damning essays on the subject in the *Humanist*, which mishandled some of the statistical fine points along the way. The Gauquelins, in turn, threatened legal action.

Kurtz struck a deal with the neo-astrologers. Their data would be submitted for a new analysis under terms agreeable to both parties. Whatever the result, Kurtz would publish the findings. According to Rawlins, he warned Kurtz that the terms of the new test would probably still yield evidence for a Mars effect. But Kurtz blew him off. And the new analysis, just as

Rawlins feared, turned out in favor of the neo-astrologers. Rawlins felt not the least bit convinced the Mars effect was real. (There was an anomaly, he felt, in the sample the statistics were based on.) But he did feel a rationalist organization should keep the agreements it makes.

Kurtz?

Not so much.

According to Rawlins, Kurtz and CSICOP engaged in a massive cover-up of the study to which they had agreed. They didn't publish the findings for two years, and when they did the result was a kind of statistical hash—attempting to explain away the Gauquelins' apparent success. Stunned, Rawlins went into overdrive, alerting other key members of CSICOP. But he found that the organization was being steered along a strictly political path: His objections were based in sound science; problem was, sound science just didn't make CSICOP look good.

Rawlins felt pressured to keep quiet, and Kurtz was in a panic over Rawlins's dissent. "I'll do anything to avoid trouble," he said, according to Rawlins.

Even more telling, when Rawlins confronted James Randi, he asked him why all the chicanery was in order. Why not simply print the study everyone agreed to and move on?

Writes Rawlins, "The reply was ever the same: We can't let the mystics rejoice."

This isn't even a remotely scientific motivation, and it reveals the extent to which CSICOP was more devoted to spreading a worldview, like religious leaders, than practicing scientific methods. Randi even spoke to Rawlins in language that likened CSICOP to a cult. "Drink the Kool-Aid, Dennis," Rawlins says he chided him, referring to the Jonestown, Guyana, "massacre" in 1978, in which nine hundred cult members killed themselves by drinking poisoned Kool-Aid.

After Rawlins forced this debacle to a head, the Committee for Scientific Investigation of Claims of the Paranormal voted to discontinue any further scientific investigations. From that point forward, they have been literally just a debunking and propaganda society. But their methods, and their unofficial mantra—*We can't let the mystics rejoice*—too often seem to describe the modern skeptical movement. For decades, in fact, the CSICOP crowd has been the public face of skepticism—names like Philip Klass, James McGaha, and of course James Randi popping up on our TV screens whenever something needs explaining.

The trouble for those of us who come in contact with a paranormal claim in the media is that this overzealous gang of self-proclaimed rationalists dominates one-half of the conversation. They are usually contextualized as the people who coldly appraise the data. But the skeptic, too, comes in bearing a bias that can render her or him just as untrustworthy as any street-corner psychic. In fact, in the wake of Rawlins's charges, other analyses appeared that backed up his account. The most damning was that of another former CSICOP fellow, Richard Kammann, who argued that the same psychological dynamic that accounts for irrational belief was operating in the skeptics by his side: "A process of *subjective validation* took over," he writes, "which I have outlined in *The Psychology of the Psychic* to account for the existence of false beliefs in the face of contrary evidence. The model says that once a belief or expectation is found, especially one that resolves uncomfortable uncertainty, it biases the observer to notice new information that confirms the belief, and to discount evidence to the contrary. This self-perpetuating mechanism consolidates the original error and builds up an overconfidence in which the arguments of opponents are seen as too fragmentary to undo the adopted belief."

Right out of the gate, it seems, the skeptical movement had run into the barriers inside their own heads. And don't worry, we'll get to the psychics a bit later on that very same score. But for now, I want to travel a bit deeper into the skeptical rabbit hole.

In the early 1970s the famous Israeli magician-cum-paranormalist Uri Geller was winning frenzied headlines. He made numerous media appearances, seemingly bending spoons with the power of his mind and perceiving information at a distance. Subsequent research into his abilities conducted and filmed at the Stanford Research Institute (SRI) yielded mixed results. He could *not* bend spoons under controlled conditions. He *did* score far above chance in remote viewing experiments. James Randi subsequently investigated, and in his 1982 book *Flim-Flam!,* he argues that films made of the Geller experiments were not shot, as had been claimed, by famous *Life* war photographer Zev Pressman. The films credit Pressman as cameraman without his knowledge or permission, claimed Randi. Even more alarming, Pressman supposedly told others at SRI that the successful Geller tests were conducted after he had gone home for the day. In sum, writes Randi, "[Pressman] knew nothing about most of what appeared under his name, and he disagreed with the part that he did know about"—namely, that Geller displayed any psychic functioning at all.

These would have been damning facts—if they were true. Pressman's own account was subsequently captured by writer Guy Lyon Playfair: "The 'revelations' [Randi] attributes to me are pure fiction." According to writer Jonathan Margolis, Pressman maintained that the videos were authentic throughout his life. Randi further quotes a physicist named Arthur Hebard accusing the SRI scientists of lying about positive results of paranormal experiments he himself had witnessed. But, well, he *also* denied Randi's statements. He told Paul H. Smith, the remote viewing Texan I met when I first arrived in Seattle, that he never would have accused anyone of lying. (Smith further explains that Hebard's account of what happened at the experiment better matches the parapsychologists' report than it does Randi's.) We are left, then, not only questioning Geller but also his chief critic. Randi, however, is far from the only skeptic whose motivations require investigation.

Susan Blackmore is something of a lioness in skeptical circles, a prize not only because she writes with grace and wit but because of her compelling autobiography. Dear Susan, it seems, was once one of *them*—a paranormalist. She has claimed she wanted to be a "famous parapsychologist." And in her essay, "The Elusive Open Mind: Ten Years of Negative Research in Parapsychology," she writes convincingly of her frustrating inability to demonstrate a psi effect. By her own account, her failed experiments convinced her that the skeptics were right: psi is the ghost in the machine—and no, ghosts don't exist either.

That really is a compelling autobiography. But her tale is not without its problems. For starters, researcher Rick E. Berger went looking through Blackmore's work and found the ten years of research she claimed in her article was actually conducted in just two or three years. Further, when her entire body of research was taken into account, she had found a demonstrable psi effect, with odds against chance of 20,000 to 1. Berger found that Blackmore had thrown out her own positive studies, ostensibly for shoddy methodology, but kept her negative results even when they demonstrated the same control problems.

To be clear, Berger and Blackmore both contend her work was so poorly designed as to be worthless in drawing a conclusion about the existence of psi (Blackmore differs with him on several of his other criticisms, too). But I do think we gain great psychological insight into the believer/skeptic dynamic by considering just what in hell Blackmore must have been thinking all these years. Why would she disregard her own positive results? Why wouldn't she simply improve her controls and keep working?

There may be a clue in her desire to be "famous." I met no stars in Seattle. I was one of a small handful of people in attendance without a direct professional connection to the field. Many of the presentations were attended by just two dozen people. By contrast, Randi's annual Amazing Meeting, the kind of skeptic's fest at which Blackmore is a regular, draws a huge crowd of bloggers, writers, and podcasters—and their fans. The July 2010 Amazing Meeting drew more than one thousand people, roughly ten times the number that attended the Parapsychological Association's tepid fête. The woman who wanted to be a famous parapsychologist may simply have found the going too hard, the prospects of becoming famous too remote. In a 1995 *Skeptical Inquirer* article she states, "I am skeptical because believing in psi does not get me anywhere." She goes on to argue that believing in psi doesn't help her understand the universe or other mysteries, but I wonder: cognitive psychologists operate on the premise that the behaviors we engage in, the beliefs we hold, all provide some kind of payoff—be it financial, physical, or emotional. From this point of view, Blackmore's famous conversion to skepticism may simply have offered her the payoffs she desired. Skepticism got her to where she wanted to be.

Richard Wiseman is another skeptical luminary. He goes out of his way, at times, to be fair—admitting some success in psi research, for instance. But he has also been accused of distorting the subject. In the episode I find most intriguing, he followed up on research conducted by Harvard-educated historian and biochemist Rupert Sheldrake. The topic was, of all things, a dog. Sheldrake had recently conducted experiments in which a dog, Jaytee, seemed to know precisely when its owner would be coming home. Again and again, in randomized experiments, just after its owner started back to the house, the dog would stop whatever it was doing and go sit by the window. Wiseman subsequently conducted a smaller series of trials with Jaytee. His data proved almost identical to Sheldrake's. Yet he went on to make the argument that his own findings disproved the biochemist's claims. He admitted the statistical similarity of their findings recently, when cornered by Alex Tsakiris on his *Skeptiko* podcast.

On a personal level, I relate to Wiseman. I have a hard time believing in psychic pooches and would like to see more evidence. (My own dog always sat in the same spot right before I came home from school, too; so hey, who knows?) That said, I think that distorting the data that does exist is counterproductive. And the lesson is that, in addition to being wary of supposedly psychic dogs, we should also be skeptical of skeptics.

CSICOP changed its name in 2006, to CSI—the Committee for Skeptical Inquiry. They publicly claimed they liked the new name's brevity. But it also seems that, by dropping the word *science* from their name, they gain a freer hand to put forth a materialist philosophy without the bother of employing the scientific method. They also captured the aura of the many TV shows airing under that banner that purport to uncover the truth through *scientific* inquiry. (Ironically, every major felony trail I cover as a reporter includes some warning from the prosecutor that CSI is fiction, and such a high level of scientific evidence doesn't exist in the real world of criminal court proceedings. So maybe the skeptics' invocation of a show broadcasting an unrealistic view is spot on.) Randi remains listed as a fellow. But today he is probably most famous for running the James Randi Educational Foundation (JREF), and his promotional vehicle—the Paranormal Challenge.

For roughly forty years, Randi has offered to pay anyone who can prove a paranormal claim. He still has the money, now more than a million dollars. And to many of his admirers in the skeptical community this is clear proof that the paranormal must not exist. But the Paranormal Challenge, too, is more sideshow than science.

For starters, the testing protocols Randi agrees to require evidence far beyond what a scientist would demand to admit a significant positive result.

Brian Josephson, the Nobel Prize–winning physicist with whom this chapter started, has noted the case of a Russian girl with a supposed gift for diagnosing medical illnesses. She agreed to the challenge and went 4 for 7 in her testing—a remarkable hit rate. Yet she was deemed a failure because the protocol demanded she get five correct.

"The case of the Russian girl was unfortunate," admits Chris French, perhaps the most level-headed skeptic on the scene. "The skeptics came off looking rather bad there. Her results were remarkable enough that they should have said, 'Well, whatever the protocol, we should take another look at her.' "

The problem is that French is speaking as a reasonable man who believes the girl would have failed by a wider, less disputable margin in a second go. Randi isn't operating from that kind of mindset. He even involved himself in the Jaytee research, claiming, among other things, that the JREF had investigated dog telepathy and found it to be false. Sheldrake said he then asked for copies of Randi's research but was told any records were informal and now, um, gone. Not exactly science.

In 2005, University of Arizona parapsychologist Gary Schwartz claimed that Randi asked him to submit his raw data, which demonstrated a psi effect, for an independent analysis. According to Schwartz, Randi listed psi-sympathetic researcher Stanley Krippner as one of the four reviewers he had lined up. But when Schwartz tried to verify Krippner's participation directly, Krippner told him the opposite: he had refused Randi's request to serve on the independent review board.

It seems to me that any truly rational group would not have James Randi as a member. But there is something very revealing in the Randi saga: that the little magician has been *allowed* to behave this way by fellow professional skeptics and the crowd of rationalists that follow their work is a testament to one thing only: our common humanity. Skeptics, after all, are believers, too. But believing in the power of science isn't actually doing science. The scientific method is itself impartial—but it is no antidote for the very human frailty of ignoring the information we don't like and embracing the information we do.

Gallup polls in 2001 and 2005 show paranormal belief has generally trended up in recent years, even as Randi has been casting spells for rational thinking. And a recent article in *Current Research in Social Psychology*, titled "Social Influences on Paranormal Belief," found that when participants were told mainstream science rejects a claim, they became more likely to accept the claim as true! "This finding ran counter to our expectations," write the authors, "but is consistent with findings that trust in science is decreasing."

Randi and the skeptics make a living posing as the figures who protect us all from irrationality and unreason. But they don't seem to have done the job. And there is nothing at all scientific and even less that is reasonable about distorting or ignoring data. Besides, when it comes to telepathy, the skeptics seem particularly extreme—because what they have been "protecting" us from, all these years, is in fact laughably small.

||||||||||||

PERHAPS THE MOST TELLING presentation at the Seattle conference occurred near the end.

Charles Tart, an academic psychologist, received a career achievement award. He walked up to the podium to speak, smiling. But it seemed a number of members preferred he not show up at all. Tart's latest book, *The End of Materialism,* is titled in a way seemingly designed to piss off the skeptics. But in Seattle he seemed bent on pissing off his colleagues in parapsychology, too—and for the very same reasons.

A long, lean, boyish, and playful kid in his early seventies, Tart chuckled his way through a talk that had his colleagues shifting uncomfortably in their seats. "We believe in God," he told them. "The skeptics have found us out, and we can't fool 'em so we might as well admit it."

Tart intended this as a laugh line, and in a sense it worked. *He* laughed. But the fact, and he knows it, is that modern parapsychologists don't fit his description. Less than a third of the respondents to my recasting of the Allison survey, for instance, "agreed strongly" with the statement, "The results of parapsychological research clearly indicate that there is a nonmaterial basis of life or thought."

Fewer than a quarter strongly agreed with the statement, "Consciousness continues in some form, after death, and includes memory and the retention of a sense of self."

Parapsychologists are so utterly *not* New Age adherents that Tart, with his deep and abiding interest in human spirituality, is a kind of walking oxymoron: an honored pariah among his own people. "I think I'm tolerated," he told me later. The takeaway from Tart's performance was, for me, simply how broadly and how dramatically the public has been lied to. The paranormal is heaped into one great manure pile. But the parapsychologists I met in Seattle not only can't be equated with the philosophers of the New Age, they seem deeply suspicious of anyone who brings up questions of spirituality at all.

I think this might be because parapsychologists are discussing a very small tear in the accepted model of the fabric of reality—one that could possibly admit all manner of New Age hoo ha, but then again might not. Theoretically, if psi exists, we cannot possibly know how much its existence will change our view of the world until we understand the mechanism by which it works. In other words, psi might not be so extraordinary a claim after all.

Want some evidence for that?

Well, for one thing, from a practical perspective, psi has little impact. What parapsychologists have found is that evidence of psi is only elicited over a vast number of trials, and even then it's best seen statistically. Run the Ganzfeld or a remote viewing test over hundreds of trials and the hit rate finally reaches what is known as "statistical significance," when chance seems a far less likely explanation than psi. In simpler terms, that might mean a 32 percent hit rate, when guessing would yield the correct answer 25 percent of the time. Over enough trials, even a 2 or 3 percent margin reaches statistical significance. And viewed in purely statistical terms, the evidence for psi is robust. One of the most compelling studies I've seen demonstrated that once enough trials are conducted, the presence of some form of mental telepathy pops out more convincingly than the evidence supporting the use of aspirin for heart attack. But that doesn't mean we can all stroll out to the nearest street corner psychic or big-ticket professional like Allison DuBois. In fact, it means just the opposite. Psychic functioning is too unreliable to be commodified or counted upon—if it exists, it really is the ghost in the machine, the phenomenon that only makes itself known at the margins.

And so this begs a question: if psychic functioning is real, but seemingly not a practical factor in our day-to-day lives, is that really such an extraordinary claim? Isn't psi rejected so hotly mostly because of its long-running connection to mysticism, its considerable buildup of Paranormal Taint? I asked the skeptic, French, the same question—with a bit more humor. "Look," I asked, "if the research indicates some small psi effect today, and rockets take off according to the laws of Newtonian physics today, if we simply admit this telepathy research into the canon—won't a rocket still take off tomorrow?"

French immediately chuckled. "Yes, yes," he said.

"Then how radical an idea is it, really?" I continued. "I mean, we won't necessarily have to rewrite the laws of physics, but more likely amend them. Isn't this usual skeptical argument, that acknowledging telepathy will force us to rewrite the laws of physics, simply a kind of reverse superstition? A kind of hysteria?"

"Yes, well," he said, chuckling louder still. "I think you might be on to something there."

||||||||||||

IN MY OWN CURIOSITY on this topic, I have consulted a few psychics over the years. And two experiences stand out.

In the first one, I see a sign for a fortuneteller on a street in midtown New York. This isn't, of course, the kind of high-end psychic we see on TV who charges hundreds of dollars an hour for a private consultation. But it is the kind of psychic we personally encounter most frequently, and I want to see what she is all about. So I walk inside, happy in summer to have found an air-conditioned room. The supposed psychic points me to a seat and sits down herself. She is young, so young in fact that I wonder how she is supposed to pick up deep, hidden information about my life when she probably at this stage knows so very little about her own. But no matter, this roughly eighteen-year-old girl of indeterminate Middle Eastern origin, asks for and receives ten of my dollars. Now she stares deeply into my eyes.

"What is your name?" she asks.

I had hoped she might predict it, so we're off to a bit of a rocky start. "Steven," I tell her.

"Steven," she says. "Steven, Steven, Steven!"

She closes her eyes and reaches for my hand across the small table between us, expertly avoiding the crystal ball.

"Steven, give me your left hand, Steven!" she says.

I give her my left hand and she clutches it in both of hers, opening her eyes again and staring at me intently. "Steven," she says, manipulating the fingers of my ringless left hand between her palms and closing her eyes again.

"You are not married!" she says.

She exclaims this as if this is psychic information, though she no doubt gained it from the sense of . . . touch. She opens her eyes again.

"Would you like to be?" she asks.

"Sure," I say.

Her eyes snap shut again. "Steven! Steven! I see love coming for you!" she says. "But Steven our time is up. For ten dollars more, I can tell you about this love I see."

This is a classic scam. Executed quickly and inartfully. And if I had proven particularly gullible, this short grifting exercise might even have turned into a long con. Disreputable psychics will tell the most vulnerable clients they are under some sort of curse—then offer to remove it for a fee.

For me, however, the experience that pushed me away from psychics was more personal than this—and it came in the past year.

Doreen Molloy is very different than the street corner psychic I saw in New York. Molloy has been vetted in double- and triple-blind readings by the Windbridge Institute, a research group that has focused on mediumship.

In these trials, Molloy proved remarkably successful and attained the enigmatic status of a Windbridge certified "Level Five Medium." I found this a bit funny, but it also intrigued me. The idea of a so-called expert medium underscored the difference between the science of parapsychology conducted by someone like Dean Radin and the performance of a psychic, who purports to be the one delivering accurate information as a professional. The imperatives are entirely different. Someone like Radin, testing psychic ability in the general population, can afford to be happy about a small but statistically significant result. Someone like Molloy needs to deliver, well, a bit of razzle-dazzle to keep customers coming back.

I took a few minutes before the call to write down the names of five people I've lost, including my oldest brother, my brother-in-law, my mother, and two friends who had died, far too young, in the previous year. I scanned over their names, feeling silly about it, but still hoping Molloy might be real, the afterlife might exist, and one or more of them might come through.

The first half-hour was entirely unremarkable, not really any different than any other reading I've had. She supplied such a wide variety of information that I could have applied it, or not applied it, to any number of people. This was remarkably frustrating, especially because early on she mentioned that an older woman, who died of cancer, had stepped forward. This woman was so weak at the end that dying was a relief.

Surely, this could have been my mother. The problem is that it also could have been millions of other mothers—and to make matters worse, Molloy had nothing with which to follow it up, nothing that was specific *only* to my mother. When I asked her for something I could use as confirmation, she asked if my mother was a musician or had a career in music.

"No," I said.

And Molloy had nothing more to give. In fact, she suggested I ask my father, who is still alive, if maybe my mother had some deeper interest or expertise in music than I had been aware. As if I didn't know my own mother, her overly sentimental taste in music and her off-key warbling whenever I heard it.

To be fair, there was one point at which Molloy went on something of a tear: I will not mention the name of the friend she seemed to be describing.

He was an avowed atheist and materialist, and in this funny world we live in, his wife might find it insulting to his memory if I publicly suggested he might still be alive. In any event, Molloy claimed another spirit was coming through and immediately scored a number of hits: "There is someone to your side," she said. "He wants to thank you. You said something for him, on his behalf, after he died, and he was very impressed. And grateful. He wants you to know he was honored. And he's also telling me he was honored . . . multiple times? There was more than one ceremony for him?"

She had my attention. This was all true.

"He was surprised today. You asked for him, right? He was surprised by that and honored by that."

I had. His was one of the five names I had written down. And I felt somehow odd doing it. We had never really been close, never socialized. But I always felt we were somehow brothers. And I deeply admired and respected this man. When he died, I asked to be included among the speakers. I even stayed up till 3:00 A.M. the night before, working for several hours to find the right thing to say.

"I'm getting that he died of cancer," she said. "It caused him problems with his digestion, and was rare. And he was very, very weak at the end and ready to go."

Yes. True as well.

"He was a writer? I'm getting that he was a writer and he is still proud of that, over there on the other side. He's saying 'I really was a good writer.' . . . I like this guy! I'm getting a lot of energy from him!"

I allowed myself one question: "Is there anything he is surprised by?"

"Over there?" she asked "Or over here?"

"Well, the question I intended was over there," I said. "But I'll listen to whatever."

"He's laughing," she reported. "He says you've got him! He was a big skeptic? A nonbeliever who thought, like, 'You die and that's the end?'"

"Yes," I said.

"Well, he was quite surprised," she said. "But he's glad with how it's all turned out."

In this entire section of the reading, which went on just a little longer, I took notes and reviewed them later. She did get a couple of details wrong. She said, for instance, that he fondly remembered a road trip with a specific friend. But I checked with the friend, who remembered no road trip at all. After reviewing my notes, I spent an hour on the Internet, trying to see if there was some way she might have connected myself, this deceased writer, and the fact that I spoke at one of his memorials. There had been no news accounts. Was there some reference to that event floating around on someone's blog?

I found nothing. And I think, to a real believer, this was it. Not just a dazzle shot, but confirmation. A conversation with someone who died, conducted through the intercession of a medium. To a skeptic, this was all just a bunch of crap—a run of lucky guesses that turned out to be right just by coincidence. And me? When I look at Molloy's performance in its totality, I am wholly unimpressed. Still, I think the following exercise is helpful. I think the most flattering claim that might be made for Molloy is that she *might* have heard from my friend. You can apply whatever percentage you like to that "might," from .0001 percent to 99 percent. It doesn't matter to me because all those numbers feel the same: rawly painful.

Might is a painful state of affairs for all of us, whether we are believers or skeptics. *Might* causes anxiety. And as a result, uncertainty is not the option we normally choose.

Instead, we choose a side:

Doreen Molloy connected me to my friend.

Or

Doreen Molloy was just guessing.

I think, in the end, when it comes to the broader picture of psi, what we have learned is more complicated than that. What we have learned is that believers and skeptics both have findings to support their cause. The believers have reams of research data suggesting some form of psi appears to be at work. But the skeptics can argue that fools and their money are soon parted—the purported psi effect so weak we can't count on it to produce accurate information on any practical basis.

And that's the stunner: even if psi is real, there isn't anything to *do* with it—at least not as the matter stands right now. The Army discontinued its remote viewing experiments because psi couldn't be operationalized to provide consistently actionable intelligence. For several years, Sony had a lab devoted to psi research that closed because they couldn't turn it into a product.

The science is clear: if psi exists, the most reliable way to see it is through the lens of statistics. As a result, believers should stop and think before ever consulting a psychic. But toward the end of this book, we'll find more reason to take heart. Because there is a better way to feel as if we've communicated with someone who's died. And for the people who experience that sense of connection, there is no better way to get through the night.

Out of Our Heads? Off with Their Heads!

Consciousness Outside the Brain

What is a mind? Do minds really exist? . . . Are minds subject to the laws of physics? What, indeed, are the laws of physics? To ask for definitive answers to such grandiose questions would, of course, be a tall order. Such answers I cannot provide: nor can anyone else, though some may try to impress us with their guesses.

—Roger Penrose, *The Emperor's New Mind*

Dr. Stuart Hameroff had a plane to catch. But he had been invited to speak at this event, and he wanted his opportunity to talk even more once it seemed like he might never go on. The occasion was the Beyond Belief Conference, a gathering peopled by the world's leading atheist scientists and philosophers—including Richard Dawkins, Sam Harris, and Patricia Churchland. The evening's moderator and event organizer was Roger Bingham, a researcher in the field of evolutionary neuroscience, who had invited Hameroff to speak after the two had dinner and exchanged emails. Bingham had a rather mercurial view of how to run the weekend-long conference, thinking it should unfold spontaneously: Consequently, he told Hameroff and all the other presenters what *day* they would speak, but not the hour. Bingham would call them up to the stage when the moment felt right.

Hameroff runs a highly esteemed conference himself, on the subject of human consciousness. He thought that improvising the order of speakers seemed a hippie-dippy way of doing things. He agreed to appear but informed Bingham that he would have to speak on day one. He had to fly back home to Arizona that night, where he had a shift to work in his capacity as an anesthesiologist.

As the first day of the conference passed, Hameroff sat listening not just to his fellow presenters but the clock as it ticked. Nearly the whole day went by with no signal from his host. So as the last session of the day approached, he cornered Bingham. "Roger," he said, "I gotta speak in the next session or I'm leaving."

Hameroff thought this reminder would settle the matter, but Bingham wouldn't commit. In fact, he seemed willing to just let the clock on Hameroff's availability run out. "Listen," said Hameroff, believing he needed to muscle his way on. "You got me here, and I'm going to speak today."

Some people, particularly in Hameroff's position, might have been more than willing to just go home. This was not a friendly crowd for him. In fact, it was the opposite. Hameroff was by now famous for having proposed, in partnership with the legendary mathematical physicist Roger Penrose, a quantum-based theory of human consciousness. The theory mounted a kind of frontal assault on all that mainstream science holds dear, insisting that consciousness originates below the level of the neuron. Even more dramatically, in Hameroff's own public estimation, quantum consciousness might not only open the door to multiple paranormal claims, but invite them in for tea.

The speakers before him had delighted in ridiculing all things paranormal and supernatural. But Hameroff was glad for the chance to confront his critics. And when Bingham relented and put him on, Hameroff made a show of flouting the differences between himself and his audience. "So far," he said, "to my way of thinking, most of this conference has been like the Spanish Inquisition but in reverse. The scientists are kind of frying the people who believe in religion. I personally am not fond of organized religion. I don't have a problem with bashing organized religion. However, I think there is room in science for spirituality."

From there, Hameroff's talk spun back and forth through the centuries. He argued that fundamental ethical and spiritual values had perhaps been embedded into the fabric of the universe. These ideal forms were located at the quantum scale, he theorized. And essentially, the human mind's apprehension of God or spirituality reflects this innate system of values. In less than thirty minutes, Hameroff connected Plato's theory of ideal forms, quantum mechanics (QM), anesthesia, neuroscience, and spirituality. Whatever points he lost for lack of detail, he gained in sheer audaciousness. But when he finished, the sound of maybe six or eight hands clapping was drowned out by the multitude's collective silence—the hushed whisper of fabric, of one hundred butts shifting uncomfortably in their seats.

Bingham was the next to speak into a working microphone. And in his deep, British accent, he cheerily noted the obvious: "You seem to have enraged just about everybody," he said.

Hameroff, in turn, smiled broadly. "That was my plan," he announced.

Looking over the audience, in fact, Hameroff could see amygdalas working overtime: He says Richard Dawkins, the King of Reason, seemed so filled with emotion that he could not even look at the podium where Hameroff stood. Instead, he stared off sulkily in the other direction. And during the next fifteen minutes or so, during the question and answer portion of his presentation, Hameroff collected the great wages of his sin. Physicist Lawrence Krauss accused him of getting all of his facts about quantum mechanics wrong. "*Everything* you said is nonsense," he announced. A philosopher who specializes in Plato accused him of arguing for something that, well, she had never read before. (Hameroff figured she should read more.) But the most telling exchange came near the end, when neurobiologist Terrence Sejnowski admitted that, in a sense, the pursuit of science hasn't yet fully explained scien*tists*. "Where I think reason and rationality really shine," said Sejnowski, "is in being able to go step by step through the evidence that you have, putting together rational arguments and writing a paper. But is that actually how we make discoveries? Right? . . . We all know, working scientists know, that it's often an accident . . . or often it's an idea you have in the shower. Where do these ideas come from? They're not deductive. From a lot of experiments and experience, something clicks."

Hameroff's body language marked the occasion: he leaned over the podium and sagely laid his right index finger over his right cheek. He rested his chin on the pedestal of his thumb and arched his eyebrows, inadvertently mimicking the state Sejnowski was talking about—the state of sudden insight. This was, at least metaphorically, what Hameroff's entire talk had addressed—the apprehension of knowledge buried in the fabric of reality itself. Merely acknowledging a mystery we all share in common seemed to open the door, just a crack, to a sense of shared experience and purpose, because for a moment or two the entire crowd seemed to fall into a kind of reverie. But then Hameroff mentioned he had a plane to catch, and the crowd, perhaps sensing its opportunity slip away, happily went back to jeering him.

The door, so briefly open, slammed shut.

The sense of unity, so narrowly present, dissipated.

Hameroff left, got on a jet plane.

His talk may not have revealed the truth about the universe. But the audience's reaction to it most definitely revealed the truth about all of us.

||||||||||||

HERE, AT THE UNIVERSITY of Arizona Medical Center, Hameroff's colleagues sometimes call him Stuart. But they seem to revel in calling him "Hammer."

In person, Hameroff is (usually) mild mannered and dryly humorous, built strong and low to the ground like a nature-made brick layer. His evident good health is the product of clean living, in Arizona dry heat. Though he is famous for his theory on consciousness, he spends his work days in the windowless bowels of a hospital surgery ward, an anesthesiologist in charge of a small troop of residents. His "Hammer" nickname arises, I suspect, from a number of areas. He himself admits he likes to engage in public debate, as he did at the Beyond Belief Conference, and he might occasionally go too far in political arguments with some of the hospital's more numerous conservatives. Hameroff's job is also, in a sense, to *drop* the hammer. He puts people to sleep for surgery, and if all goes well he wakes them up again. Anesthesiology is perhaps the least glorified aspect of medicine. The surgeon, so skillful in rending one millimeter of flesh from another, gets the glory. The anesthesiologist, twirling the knobs that control a collection of gases, goes relatively unnoticed. "It's like flying a plane," says Hameroff, waiting for his next surgery in a set of scrubs. "Hours of boredom punctuated by moments of sheer terror."

He is only half-joking. When things go smoothly, when the patient's blood gases and vital signs are all as they are supposed to be, the anesthesiologist can stay the course till something changes. But when a patient goes south, it's "time for a new plan," as Hameroff puts it. Hameroff, however, doesn't have a worrier's personality. With big, alert almond eyes and a face that settles into a smile by default, Hameroff has the bravado of a pilot—a man who steadily lifts a multitude of souls into the ether, then settles them down again, safe to get on with the rest of their lives. In his early sixties, he also has the air of a man who has lived long enough to know himself and be comfortable with what he sees. He spends his morning stopping in to see the patients he'll put to sleep later, leaning over their bedsides and pelting them with a steady stream of humor only an anesthesiologist would find funny. "Don't worry," he tells patients, "the whole operation will only take a second, as far as you know."

"Hey," he says, "you'll be right back."

The patients, every last one, smile appreciatively. Extending his flying analogy, he likens them to nervous travelers, waiting in an airport. "This isn't easy on any of them," he says. "I'm the one standing up and I do this every day. This is a difficult experience for them."

One older woman starts to cry when he turns to leave. "Whoa, whoa," he says, taking the woman's nervous, shaking hand in his own. "What's wrong?"

"I'm scared," she says.

"I know," replies Hameroff. "I understand. But you're going to be okay. We're going to take good care of you."

The woman is having gall bladder surgery. Minor stuff. But she is scared she'll never wake up. Hameroff keeps hold of her, with both hands, and stares into her eyes. "Hey," he says, "I promise."

The woman's panicked eyes soften a little. Hameroff makes sure she has a relative to sit with her until surgery. Then he finally lets go. Seeing him in this environment is something of a revelation. Helping people negotiate the tricky boundary between life and death is difficult work on the best day. Reminders of his own mortality, and that of his loved ones, lie on tables all around Hameroff all day long. But publicly, at least, the fifty hours a week or more he spends working in the innards of a hospital are just his sideline—the day gig that supports his real calling: consciousness research.

Hameroff is, in many respects, an unlikely candidate for this kind of leading role in international intellectual life. He rose, seemingly in a matter of months, from an almost complete unknown in a Tucson hospital to one of the most pilloried and publicized figures in consciousness studies, invited to speak all over the world.

In ways less well known, he was truly born to all this. He hails from a colorful family in Buffalo, where his father Harry, a former carnival barker, worked the stages of burlesque theater and vaudeville, which explains Hameroff's own taste for corny comedy and his ease speaking into a microphone. Hameroff's paternal grandfather, Abraham, was also a big influence. An intellectual dilettante, Abraham knew a lot about a lot—and never saw much need to turn it into a paying job. "I owe him," says Hameroff, "because he got me involved in thinking about a lot of big ideas at a very young age."

His grandfather was supported by what Hameroff refers to as "family money." And so the old man was free to come in and give his grandson science books and engage him in conversations about Einstein, at any time of day. "It was heavy stuff," remembers Hameroff. "I think I was about six years old when he started that."

Those conversations rendered him, he says, an "idealist, which is really the research end of my business."

In elementary school, he told his science teacher he wanted his class project to address "the fundamental nature of life."

"How are you going to do that?" his teacher asked.

The young Hameroff wasn't sure. He wound up designing a more basic experiment. But the idea gestated in his subconscious mind. "I used to have a kind of nightmare," he says, "a recurring dream in which I was trying to understand the nature of reality. In the dream I was standing at the edge of the universe, and trying to get to the other side."

When the time came, Hameroff chose medical school. He was interested in the mind-body problem. How is it that electrical signals and chemical pulsations between neurons result in our experience of life? Neurology seemed a likely career path. Maybe psychiatry. Then, while Hameroff was doing an internship at the Tucson medical center, the chairman of the anesthesiology department offered some advice. "He told me that anesthesiology was key to understanding consciousness," says Hameroff, "and when I looked at it, he was right."

As Hameroff quickly learned, all the normal brain processes continue under properly administered anesthesia save one: consciousness. The patient goes right on breathing, neurons go right on firing, secreting, and absorbing chemicals. The only thing missing from "normal" brain function is the most crucial aspect of human experience: awareness. Consciousness.

Hameroff liked the idea of specializing in anesthesiology even more when he realized he would be sitting at a sweet, personal nexus: the meeting of blood, anesthesia, and neural cells isn't just medicine. It's also physics.

His grandfather.

Einstein.

His dreams of standing at the edge of the universe.

In anesthesiology, the interlocking details of Hameroff's personal narrative just clicked. But the result didn't manifest itself with any drama for another twenty years. In the interim, Hameroff got married. Had a kid. Got divorced. Life happened, but he never lost interest in the topic. In fact, he continued working on it, publishing a collection of medical papers and penning a 1987 book, *Ultimate Computing: Biomolecular Consciousness and Nanotechnology*, that went relatively unnoticed outside a small coterie of professionals interested in the topic of artificial intelligence—the problem of whether or not a computer can ever be created that is capable of mimicking all the operations of the human brain.

The book is essentially a dense, knotty, intellectual ode to the microtubule. These unbelievably tiny, 25-nanometer-wide tubes were only discovered (and accepted) in the 1960s. And Hameroff seems delighted, recounting the accidental way they blinked into view. "The solution used to affix cells to a microscope slide was changed," he says, "so at first people thought microtubules were an anomaly produced by the new solution."

In reality, the old solution had been dissolving microtubules before scientists could have the pleasure of seeing them. Eventually, scientists did accept the reality of the microtubule, which proved ubiquitous, appearing in literally every biological cell—plant or animal—on the planet. Scientists also discovered, more slowly, over decades, that the microtubule is perhaps the single most versatile biological component in nature. The way our own bones keep our flesh from piling up in a heap on the floor, microtubules are the cytoskeleton that supports the structure of every cell. But they also act as conveyor belts inside cells, moving needed chemical components from one cell to another. And they are capable of moving them*selves*, comprising the levers, so to speak, that divide chromosomes.

Hameroff has come to believe that microtubules are essential for understanding human consciousness. And by way of argument, he points to a single-celled organism, the humble paramecium, which lacks the neurons and synapses we boast in our brains. "The paramecium swims around, finds food, finds a mate, and avoids danger," says Hameroff. "But it doesn't always swim toward food or away from danger. It seems to make choices and it definitely seems to process information."

How, precisely, can a brainless cell be said to *know* when it is time to divide, eat, or mate? Hameroff says the information processing takes place in the microtubule, which gives the cell its shape *and* comprises the source of its primitive form of thinking. He published papers on these ideas but found resistance everywhere. Microtubules were initially easier to understand as simple girders, so there was an intellectual and scientific utility in denying them any further purpose. But he found the greatest resistance to his research among proponents of artificial intelligence (AI).

Supporters of hard AI say that once we have computer bits and switches processing information with the same computational power as the human brain, we'll succeed in recreating human intelligence—and consciousness. But they set their timelines for when this might be achieved and write their funding and research proposals based purely on the computational power of neurons alone. "I was telling them to push the goal posts way back," says Hameroff, "to account for all these microtubules processing information *inside* the neuron. They didn't like that."

Of course not.

No one wants to be told the holy grail of his career is at best decades further down the road. So Hameroff began collecting intellectual enemies early in his career. *Ultimate Computing* was his cannon shot at their broadsides. Hameroff admits that, for a time, he puffed himself up with the idea that the sheer, extra computing power suggested by microtubules explained consciousness. Then a friend approached him, and with the wisdom of Socrates, merely asked a probing question: "Say you're right," the friend said, "and all this processing is going on inside the microtubules. So what? How does that explain consciousness?"

To his chagrin, Hameroff realized his friend was right. He had, like the hard AI proponents themselves, merely put forth a kind of faith-based argument: if we assemble enough computing power, in the right formation, we'll get Data from *Star Trek*—a conscious machine with a sense of self and a personal narrative, with goals and wishes and the ability to appreciate, viscerally, the beauty of a well-composed sonata. The problem is that there is no understood mechanism by which our own sense of self is generated. How do the firings of different networks of neurons, how would computation produce any sensation or thought at all? Hameroff now realized he had provided no answer. Merely driving the discussion further down, inside the neuron to the microtubule, suggested how much raw computing power human beings hold—but it didn't solve the mind-body problem, the riddle of consciousness. Not even close. Thrust back into this state of wonder, into a state of open-mindedness, Hameroff found himself alone one night, several years after he published *Ultimate Computing*. He was in his early forties and living in a house in the Tucson desert, reading a book by Roger Penrose.

A British mathematical physicist, Penrose reached a new level of intellectual and popular celebrity with the book, *The Emperor's New Mind*, a bestselling yet densely scientific tome that leads readers through physics, cosmology, mathematics, and philosophy, all to arrive at the mystery of consciousness.

Interspersing pages of equations with lucid descriptions of complicated scientific concepts, Penrose overturns several apple carts at once. And Hameroff gobbled up the details.

Penrose's book argues that the difference between us and a computer is one of understanding. Computers will at times grind on forever without finding a solution. But a human mind can step back from these calculations and understand when they are headed nowhere. Most famously, Penrose builds a logical argument off the work of mathematician Kurt Gödel, which demonstrates that some mathematical statements can neither be proven nor disproven—cannot be computed—yet we as human beings can still know them to be true or untrue. This ability to grasp information that cannot be expressed computationally suggests to Penrose that there must be something other than computation at work in the human mind.

Hameroff was already transfixed. It was late at night. But he didn't want to go to bed. He felt himself alone in a pool of lamplight, exposed to some of the most important ideas he had ever read. The reductionistic view of the workings of mind and brain, so dominant throughout science, was being systematically demolished in the 500-page book in his hands. But it was toward the end of *Emperor,* when Hameroff perceived his own personal connection to the material, that he realized Penrose might *need* him. Because at the end of his book, after swimming through hundreds of pages of dense science and mathematics, Penrose the author surfaces in what seems a new world—a world, Penrose surmises, in need of a new physics.

In his closing pages, Penrose speculates that the key to understanding consciousness lies in quantum mechanics. In his view, the Newtonian worldview of *deterministic* physics isn't sufficient to explain consciousness as the pinging of neurons, equivalent to the bits and switches of a computer carrying out computations. So perhaps, he suggests, some clue to the answer lies in the more complicated, *indeterminate* nature of quantum mechanics. In fact, Penrose argues that human thinking speaks to the existence of something *beyond* both classical and quantum physics, something that includes and perhaps transcends the two.

I describe the quantum in greater detail later, but for now, understand this: Penrose knew he lacked evidence for the quantum-based part of his thesis. In fact, the bulk of then-current evidence—or at least, then-current scientific thinking—suggested he was wrong. Given this conflict, what area of the brain might be able to house the tenuous, fragile operations of the quantum?

At the end of his book, Penrose flatly admits he isn't sure. But he extends a bony finger down into the subatomic realm just the same, saying, *Here is where the answer must lie.* And Hameroff, sitting alone in his house, looked at where Penrose was pointing and realized he had in fact already been there—through his research into the microtubule. Sitting there that night, he realized that he—an anesthesiologist sitting up too late, reading in his desert home—might be the one who could help Penrose take the next step.

||||||||||||

HAMEROFF'S LIFE MOVED AWFULLY quickly from there.

He contacted Penrose and was invited to meet with the eminent scientist on Oxford's campus. Hameroff had friends in England and got on a plane. When he finally arrived he was, like most people, immediately taken with Penrose—his impish smile and piercingly intelligent eyes. "Roger's just . . . on another level," says Hameroff.

Penrose's office was crowded with books and papers stacked up in great heaps all over the desk and floor. He turned to Hameroff casually, with the practiced air of a professor used to addressing students. Hameroff wondered if he might have flown all this way to get a quick hook.

Hameroff talked. Penrose mostly listened, asking for a point of clarification whenever he thought it necessary. Otherwise, he remained silent—his face inscrutable. Hameroff laid out the basic understood facts of microtubules and the less well-known role these tiny structures seemed to play in human consciousness. Activity in the microtubules is constant, Hameroff told him, except when a patient is under anesthesia. Anesthetic gases work by means of weak, London forces, intermolecular forces caused by quantum dynamics. This loss of awareness under anesthesia could be the key, he said, to understanding a quantum theory of consciousness.

Hameroff felt he had made his best case. But Penrose just smiled, thanked him for his time, and shook his hand.

Hameroff walked out on to the Oxford campus, thinking that was the end of his relationship with Roger Penrose. "I didn't think he was impressed with the idea at all," he says. "I mean, I took my shot and that was it."

A couple of days later, Hameroff was still in London, when a friend met him in a pub. "Have you heard," he asked, "about the talk Roger Penrose just gave?"

The scientist had apparently just delivered a lecture on *The Emperor's New Mind*. And in closing, he made an announcement that caused quite a stir in the crowd. He had just met with an American scientist named Stuart Hameroff, who had explained to him that microtubules could be the structure he was looking for to develop his quantum theory of consciousness. Penrose had liked his idea after all. And over the ensuing months the pair met, often, to refine their thinking. What they came up with would, at least in terms of the pop culture zeitgeist, capture the moment.

I promised, or perhaps threatened might be a better word, to address quantum mechanics a bit more fully in this chapter. And so I shall. Interested readers who would like to understand it more technically and in all its confusing glory should consult the Notes and Sources at the back of this book. But what readers most need to understand is this: disputes about quantum mechanics seem to revolve less around the science of QM than what that science *means*.

In other words, the scientific method has sussed out a lot of facts about the subatomic realm. But what those facts say, or don't say, about the nature of the universe and, well, reality, remains a matter of debate. And so we are in a curious position, at this time in our history, as a species: the science we believe describes the underpinnings of our world is embraced by some—New Agers, mostly—as the foundation of both matter and spirituality. Their philosophy is that the strange properties of the quantum suggest some mystical component to the nature of man. Others, mostly materialists and atheists, accept that QM describes the subatomic but deny it has any importance beyond that. *Their* philosophy is such that we can still understand our world and ourselves in the light shed by Newtonian physics—all is matter, smacking this way and that.

The result is that, by invoking the quantum realm, Hameroff and Penrose stepped right into an ongoing culture war—in which two opposing sides are claiming they know exactly what to make of the quantum, though the truth is, we can't yet claim that kind of knowledge.

In the quantum universe, particles regularly perform seemingly impossible feats: appearing in two places at once; communicating information across distances; blinking out of existence in one spot and reappearing suddenly in another.

Here are just a few of the stranger quantum findings, each of which has been confirmed by numerous scientific experiments: separate particles of matter can maintain nonlocal connections, called "entanglement." In instances in which particles have become entangled, a change to the state of one particle results in an immediate, corresponding change in the other. These connections persist across any distance, from meters to miles.

In the phenomenon of quantum tunneling, described briefly in chapter 2, a particle can pass right through a seemingly impervious barrier.

Then there is superposition, in which subatomic matter is said to exist in all its possible states at once—until it is interfered with in some way. At that point, the wave of possibilities is said to "collapse" and assume a measurable state. Even then, a precise, complete measurement remains impossible. We can know a quantum particle's position, for instance, but not its velocity. If we choose instead to measure its speed, we cannot know its place. These findings have been with us, in something approaching their modern form, ever since Max Planck and Albert Einstein first began working on the conundrum of how light can display the properties of both a particle and a wave.

Think of a particle as a billiard ball, located in a specific place and completely distinct from the other billiard balls on the table. Propelled by some outside force, the billiard balls might interact, hurtling into one another with a satisfying *thwack* and moving on in completely predictable ways. But any one billiard ball is decidedly *not* another billiard ball, and cannot be. Now conversely think of a wave as . . . a *wave*. Rather than being local to one space, a wave is spread out. Waves can in fact interfere with one another, joining and unjoining as in a roiling surf. This is the Newtonian world that we live in and observe every day. But QM has revealed an entirely different reality, in which light has proven to be both particle and wave.

In what's known as the classic, double-slit experiment, photons, electrons, or any atom-sized objects are shot at a screen, impervious but for two tiny slits. A photographic plate sits on the other side of the screen, there to record where each photon lands. Close one slit and the plate will record a logical pattern of photons, with the bulk of them cropping up right across from the open slit. Open both slits and the plate will record bands of photons landing in a pattern of varying intensity, consistent with a wave. So far, so good. Newton rules. But this is where things get well and truly weird: Because if just one photon is emitted into the apparatus at a time, while both slits remain open, scientists would expect to see two regular patterns of photons striking the photographic plate—one behind each slit. But, that isn't what happens. Instead, after shooting lots of photons, one at a time, toward the slits, we find the same interference pattern we saw when the photons were emitted in a flood. The scientific conclusion is that each photon individually travels through both slits at the same time—meaning, every individual particle behaves *as a wave*.

In any case where photons are being emitted individually, then literally speaking, wave interference cannot take place. But wave interference is precisely what is observed. The same experimental results hold true not just for photons, but neutrons, electrons, and protons as well. No one particularly *liked* this conclusion. But it was staring them in the face: photons act as a wave when they aren't being measured, and as particles when they are.

It is true that the act of measuring a quantum particle involves striking it with another particle of comparable energy and size. So it isn't just the act of *looking* at the wave function that causes it to collapse into a particle. Failure to make this distinction is responsible for much of the controversy surrounding QM. Still, even when properly understood, QM remains a challenge to our philosophy. The role of the experimenter in consciously choosing when and how to measure a particle *does* directly influence its behavior and in this sense creates a dramatic link between the person doing the observing and the quantum particle being observed. Unless you've read up on the quantum for yourself previously, and perhaps even if you have, you're feeling a bit lost right about now. In a sense, so is modern physics—lost, that is, and found. Because the strange reality of the quantum realm is precisely what underpins our own more predictable, billiard ball lives.

Physicist John Wheeler added an extra measure of crazy to our understanding of the quantum by formulating a "delayed-choice" design. In this experimental variation on the double-slit exercise described, either slit can be closed *after* the photon has passed the initial screen but before the photon reaches its final destination, where its location is recorded. The Wheeler experiments confirm a single photon will behave like a wave if both slits are allowed to remain open—a particle if one is closed. The result suggests the particle "knows," after it passes through the slits, that one of the slits will ultimately be shut. Later.

Don't even worry about the voice in your head shouting it can't be true. (That's just your amygdala talking its oft-untrustworthy paranoia.) Just accept, for now, that it is true, and that these experiments raise a host of issues, both philosophical and scientific, including the nature of the relationship between the observer and the observed. I've long accepted, as a reporter, that my presence on the scene with a notebook and a pen causes the behavior of my sources to change, often dramatically. In fact, I sometimes have to hang around my subjects for hours or even days before they begin to let down their guard and behave as they normally do. I would never expect the fundamental building blocks of reality, however, to behave in the same seemingly conscious manner as the politicians, cops, and criminals I interview. Yet the results of these experiments suggest that is precisely what happens. This aspect of quantum mechanics has spawned a host of analyses, including the Copenhagen interpretation, which claims that quantum waves of energy "choose" a particular state only under observation. The other, leading contender is the many-worlds interpretation, which gets around this idea that we somehow create reality by measuring it by arguing that all the possibilities of a wave are realized—in separate universes. As you might imagine, mystics are among those who love the Copenhagen argument, and no matter how irrational it sounds, many rationalists dig many-worlds.

I find the many-worlds interpretation most intriguing for what it says about the many scientists who in their love for this theory arguably stop practicing science at all. The idea that our universe is one of a seemingly infinite number, splitting off into still more "copies" of the universe (or perhaps just copies of quantum wave functions, depending on who's doing the talking) as different possibilities are realized—as in one universe the electron spins *up*, in another it spins *down*—is pretty tough to accept outside the Cineplex. And so the many-worlds interpretation of quantum mechanics is in a sense modern science's great Achilles' heel—the evidence of what desperation can do to us all. Because in trying to reconcile a mechanistic picture of the world with the vagaries of quantum mechanics, a surprisingly large number of scientists endorse the seemingly unscientific idea of an infinite number of parallel *some*-things operating unseen alongside ours.

I don't call the many-worlds theory unscientific just because it sounds like something out of a science fiction plot. I call it unscientific because more than fifty years after it was first introduced, by a theoretical physicist named Hugh Everett, no one has devised an experiment to test whether the many-worlds theory is so. Of course, the scien*tific* method calls for scien*tists* to construct testable hypotheses. And in fact, if a theory isn't testable it's normally not considered sound science. Some scientists rail against many-worlds for these very reasons, yet the idea has largely been granted a pass.

In a poll conducted by political scientist L. David Raub of leading cosmologists and other quantum field theorists, in fact, 58 percent responded, "Yes, I think [the] Many-Worlds Interpretation is true."

Proponents of many-worlds include some of science's greatest luminaries, like Stephen Hawking, Murray Gell-Mann, and Richard Feynman. Supporters seem to like this theory, no matter how far-out it sounds, because it allows them to go on looking at *this* universe exactly the way they already do. These many universes, the theory goes, don't interfere with each other in any way that might force us to refigure our current understanding of physics. But this lack of direct observation, from any one universe to another, is also what nudges the idea from the empirical confines of science toward philosophy.

There is some hypocrisy here: mental telepathy, which we discussed in chapter 2, is dismissed by the scientific mainstream—despite mountains of well-controlled scientific research suggesting *some*thing small but real is going on there. In contrast, an untestable idea like many-worlds, unencumbered by the stigma surrounding the paranormal, can win wide acceptance from these very same people.

That QM has driven a great many scientists into a claim that, like the supernatural, seems to lie beyond the current reach of science should not come as a surprise. The legendary physicist Richard Feynman knew what a challenge QM presented to our understanding. Often, this kind of talk is dismissed as hyperbole, scientists patting themselves on the back for understanding what mere mortals can't—or in Feynman's case, a beautiful mind speaking off the cuff. But in a philosophical sense Feynman meant every word. In a lecture collected in *The Strange Theory of Light and Matter*, Feynman says of quantum mechanics: "It is my task to convince you not to turn away because you don't understand it. . . . I don't understand it. Nobody does."

We should be clear here. In a *practical* sense, scientists *do* understand quantum physics.

Quantum mechanics enabled scientists to predict new kinds of particles, which they subsequently found. And quantum mechanics helped us discover the forces that bind atoms into molecules—in short, chemistry. Quantum mechanics, or the science of subatomic particles, is the foundation of computers, televisions, all the electronics we consume. Our knowledge of the subatomic world *is* powerful, powerful enough to make it work for us. But what we don't understand about quantum mechanics is *what it means* for our philosophy. And so the iPod has also come with a terrible cost—and I don't mean $250.

Traditionalists, including Einstein, thought the new uncertain foundation of the universe that had just been discovered was itself illusory, though he himself had helped to usher in this new understanding! If the strangeness of the quantum world was not somehow explained away, complained Einstein, "I would rather be a cobbler, or even an employee in a gaming house, than a physicist."

There is something important to consider here about the human beings who practice science. Einstein, like many physicists and rationalists, simply liked the idea of a predictable world better. He chose to enter physics because—like Hameroff, dreaming of standing at the edge of the universe—he liked the idea of being able to explain how everything worked. And he was honest enough to admit his own disappointment in the world Quantum Mechanics described. Today, physicists and cosmologists remain embroiled in a controversy over whether the Many Worlds interpretation of QM even qualifies as science. This infinite number of hypothesized universes could turn out to be real. But for now, their reality seems largely subjective, a matter of preference.

The terrible cost of QM is that it has thrown up the shutters on us all; and in its mysteries, it has forced scientists to go beyond the evidence and data at hand to find some way of claiming the world they prefer is also the world that is so. This is the stormy, conflicted milieu into which Hameroff and Penrose reached to explain consciousness. In doing so, they immediately subjected themselves not just to scientific criticism—but to scientific anger. That is in great part why Hameroff got the reception he did at the Beyond Belief conference—a reception that was at least as emotional as it was reasonable. And that is why so many scientists continue to castigate his quantum-based theory of consciousness, even though we have more reason than ever to believe QM might play some role in biological systems.

|||||||||||

HAMEROFF REMAINS PERHAPS BEST known for his appearance in the surprise hit documentary film, *What the Bleep Do We Know?,* a New Agey treatment of quantum mechanics in which human beings are said to create their own reality. Critics of the 2004 film contend it got its science something less than half-right—stretching the role of the observer in choosing whether to measure a particle into an almost Godlike power to succeed in life, love, and career merely by thinking it so. The movie has since become a kind of Exhibit A in arguments skeptics make against the invocation of what is often dubbed "quantum magic" or "quantum quackery." But Hameroff, who was one of numerous talking heads in the film, isn't afraid to speak up for it—to a point. "I thought the movie was meant to entertain and inspire," he says, "and on that score it succeeded. Look at how popular it is. I stand by everything I said in the film, and as for the rest, I had nothing to do with it."

The things Hameroff says in and outside the film, however, are more than enough to earn him a spot in the skeptic's Hall of Shame. Telepathy, the afterlife, spirituality: Hameroff has claimed that a quantum origin for consciousness could open the door to all these things. Entanglement might explain telepathy and contribute to an afterlife spent roaming the cosmos in a tiny subatomic form. Crazy stuff, that. But it's safe to say, more than fifteen years after they first introduced their theory to the world, the jeering is loud but the jury is still out on the Penrose-Hameroff model of consciousness.

In sum, what became known as Orch-OR, or orchestrated objective reduction, would explain consciousness *and* the primary mystery of quantum mechanics in one shot. The observer effect, in which the waveform is said to "collapse" into a particle state, *is* consciousness; each conscious moment is a collapse. The particulars of how this happens are as complicated as you might wish them to be, revolving around a theory of quantum gravity. But essentially, the Penrose-Hameroff model relates collapse of the wave function/consciousness to fundamental components of the universe—like the properties of space and time. They cannot be explained or reduced because there is nothing to reduce them to.

The most often cited argument against Orch-OR was given early on by physicist Max Tegmark, who argued that microtubules could not sustain quantum states for a long enough period of time to be relevant to neural processing. But Hameroff and a physicist named Jack Tuszynski countered by saying that Tegmark had improperly modeled the Orch-OR theory, rendering his calculations inaccurate. The state of play hasn't really changed since then, and so, in short, no one has yet falsified Orch-OR or completed an experiment that suggests Orch-OR must be so. For now, the theory's real value lies in the reaction it has provoked, demonstrating how desperate believers are to earn the scientific validation of quantum physics, and how deathly afraid materialists are of considering that quantum mechanics might validate some paranormal claims about the nature of the world or even influence our philosophy.

In fact, the most telling assault on Orch-OR was the one launched by philosopher Patricia Churchland *before* the Penrose-Hameroff model was even published. "She couldn't wait until it even came out to attack it," Hameroff told me, smiling. "I mean—what is that?"

Churchland is herself renowned. But in this instance, her emotions seem to have got the better of her. She tried a twin-barreled, microtubule/anesthesia-based argument on Hameroff, an expert on microtubules and anesthesia. Perhaps predictably, Hameroff's response demonstrated the errors she and coauthor Rick Grush made in their haste to object to Orch-OR. Still, I can't help but admire Tegmark and Churchland (and anyone who has bothered to do any research in association with shooting down Hameroff's idea). More often his critics do . . . nothing.

As a kind of pushback to the more whacky ideas in *What the Bleep Do We Know?,* a lot of popular science journals and a lot of popular scien*tists* argue that there really isn't anything crazy about quantum mechanics—and all arguments that QM plays a role in consciousness are de facto bunk. All this stuff going on at the micro-level of reality is only going on there, anyway, so in effect, *Who the bleep cares?*

Quantum mechanics is the science of the small. How small, you ask? Well, as physicist Brian Greene writes in *The Fabric of the Cosmos,* at the infinitesimal level of the Planck Scale, where the weirdest aspects of the quantum take hold, "there would be no such thing as a distance shorter than the Planck length." So you take the skeptics' point. Quantum effects may underlie our physical world, but they are too small to influence what we perceive.

As we scale up from the quantum micro-world of subatomic particles to the macroworld of people and things, all the weirdness of the quantum literally disappears and has no effect on the reality we interact with every day.

This wall, between the big and the small, has for a long time seemed particularly high and sound. But as I stated earlier, the authority of science is its method—not its current base of knowledge. And the fact is, the "too small" argument no longer works so well.

The micro-scale of the quantum interacts with the macroworld of you and me far more than we knew just three years ago. These new findings may not have made the Penrose-Hameroff model *likely,* but they do render the idea that quantum effects play a significant role in consciousness more plausible. Biologists are in fact finding that quantum processes may underlie the migratory habits of birds, the human sense of smell, and photosynthesis in plants. Quantum processes are proving far more stable and resistant to warm, wet biological environments than anyone ever thought possible.

Experiments are also seeking quantum processes in human visual perception, which should perhaps not come as such a great surprise. Experts in art, architecture, and music have long known about the golden ratio—a precise mathematical formulation that holds mysterious aesthetic appeal. Why is the Mona Lisa such a compelling figure? After all, *M'lady's forehead is so wide!* But the *Mona Lisa* is so appealing because her proportions are equal to the golden ratio, which can be expressed in a variety of forms or as a mathematical figure, 1.6180339887. Mona's mischievous eye and the corner of her playfully curled lip are lined up in so-called golden sections within the painting. And what does this have to do with quantum mechanics?

Well, in early 2010, experimenters at Oxford University found that quantum particles in a resonant state reflected the same ratio: 1.618. The lead investigator, Radu Coldea, claimed the ratio is too precise for its appearance in quantum physics, and art, to be a coincidence. "It reflects a beautiful property of the quantum system—a hidden symmetry. Actually quite a special one called E8 by mathematicians, and this is its first observation in a material."

Was Hameroff's Beyond Belief presentation exactly right when he talked, years before this experiment, about perfect forms being embedded as a fundamental property of the universe? Is this, in fact, another sign of an intimate relationship between the macro- and micro-scales? As Penrose noted in *Emperor's New Mind,* two experiments, including one conducted in 1941, demonstrated that the human retina can react to a single photon. This is a direct example of a physical interaction between the micro- and macro-scales, known for sixty-eight years, which raises a question I think we can answer: if that discovery about the retina being sensitive to a single photon was known in 1941 (and reconfirmed in 1979), how had this wall between micro- and macro-scales ever stood so high for so long in the first place?

Well, because we wanted it to, we *needed* it to.

Give the wall some credit. Maintaining a strict dividing line between these two worlds was and is convenient. There is a real difference between what we observe at these two scales of reality. None of us goes winking out of existence in one place and appearing in another, for instance, like quantum particles. But it's also the case that we don't know precisely where to draw the line between the micro- and macro-worlds at all—and never really did.

In a 2008 *Seed* magazine article, which I suspect will go down as one of the most important pieces of popular science journalism ever written, author Joshua Roebke captures experimental physicist Anton Zeilinger's long, dark, and productive night of the scientific soul. Zeilinger has been finding more and more quantum processes going on in *macro*-scale objects, suggesting these strange quantum occurrences are going on all the time, in objects large and small, and are simply out of the range of our ability to perceive them. Zeilinger admitted to Roebke that he was in fact so thrown by what these experiments were showing him that he felt he had no choice other than to give in to the mystery, and hire, of all things, a philosopher.

In fact, Roebke's article ends with Zeilinger placing just such an employment ad. (Can you imagine how that must have read? *Wanted: World-renowned experimental physicist seeks philosopher. Someone who can take long walks on the beach with him in consideration of what his experiments suggest about the nature of reality.*)

I also, of course, felt compelled to call Zeilinger for this book, at the lab where he works in Vienna. His press rep got back to me almost immediately. When a phone call was hastily arranged, however, it quickly became apparent that neither Zeilinger nor his press rep understood what I meant by "I am a journalist and I would like to talk to Dr. Zeilinger about what he has learned from the philosopher he hired."

"Are you a philosopher?" Zeilinger asked me. I could hear what sounded like traffic on his end of the line. And even though there was a language barrier in effect, I attempted a lame joke. "Well," I said, "we're all philosophers to some extent. But no, I'm a journalist."

"A journalist?"

"Yes," I said. "I got in touch to interview you. Did you succeed in hiring a philosopher?"

"Yes," he said. "In fact, I have consulted with a great many philosophers. But . . . you are a journalist?"

"Yes," I said.

"Well, I am busy," said Zeilinger. "I am driving. I will call you later."

I knew what that meant. And here at the macro-scale my predictions are pretty good: I never did hear from Zeilinger again. Neither he nor his press aide responded to any more of my emails. But there is enough of Zeilinger out there, on the record, for us to gain some meaningful traction. The reductionistic practice of science has in fact led Zeilinger to a strange conclusion: in short, while most of mainstream science and all the hard-core skeptics spend their time arguing that the world ultimately reduces down to physical matter, Zeilinger claims his research has demonstrated exactly the opposite. What really exists, and what seems to have properties like nonlocality, are the properties, or information, contained in the atom. This information can, in a quantum state, even be sent from one particle to another and is the basis of quantum cryptography—an end the U.S. military pursues to send super-secret messages. As Zeilinger puts it, "Matter itself is completely irrelevant. If I swap all my carbon atoms for other carbon atoms, I am still Anton Zeilinger."

This is big, important stuff, suggesting that the reductionistic practice of science has found that, at bottom, this isn't a classically material universe after all. Tough news, I should think, for materialists to take. But perhaps the most appealing thing about Zeilinger, in all his public remarks, is that he acknowledges these mysteries, notes our current inability to understand what they mean about the nature of reality, and then refuses to say more.

It isn't that Zeilinger is a shrinking violet. He met publicly with the Dalai Lama, discussing the common ground shared by Buddhism and experimental physics. And he has given a few interviews in which he seems happy to acknowledge that quantum conundrums raise profound questions about the nature of reality. He is also not alone; other modern physicists are pointing toward the quantum as being far more integral to any true picture of the world than mainstream scientists and philosophers have cared to admit.

In *My Big TOE,* or Theory of Everything, Tom Campbell, a nuclear physicist and consultant to NASA, argues that quantum mechanics forces us toward an entirely new understanding of physics, in which consciousness plays a central role. And physicists Bruce Rosenblum and Fred Kuttner make the case in *Quantum Enigma* for a physics that at least acknowledges and addresses the mysteries raised by quantum experiments. While openly deriding and mocking more outlandish theories like those in *What the Bleep?,* they bravely accuse their brethren of hiding the real mysteries raised by the quantum away—the skeleton in physics's closet. Like them, Zeilinger generally stops short of giving statements so much as illuminating the relevant questions.

I can't say I blame him. There is a point, after all, at which we drift into mere speculation, in which we are simply giving voice to our own biases, conceits, and wishes—and at that point we are certainly revealing ourselves, while quite possibly saying absolutely nothing about the world. It is this inability some of us have to shut up and admit our lack of knowledge that dominates far too much of our conversation and creates such a massive verbal train wreck out of our discussions of the paranormal. Consider this: if we again take the word *paranormal* as referring to that which science hasn't yet explained, then Anton Zeilinger's experience would suggest the ultimate nature of *reality* qualifies as paranormal.

There is nothing outrageous contained there—no reason for celebration or grief. The mystery of the world exists apart from our judgment of it. Reading the physics-specific journals and message boards, in fact, gives a clue to this. The questions they're asking are fundamental: *Does time as we know it really exist, or is our experience of it just a product of our perception? Are we living in one of many parallel worlds—or a multiverse? Do we live in a hologram? Is consciousness a fundamental property of the universe?*

These are questions that make a statement in their asking—a statement about our own status as a species just out of the trees, trapped on a dusty rock, floating through space . . . but to acknowledge what this means, that in essence we *do* live in mystery, we *do* live in a paranormal world—is verboten. The *P* word is the bearer, after all, of that unholy Taint.

At this time, I should point out there is nothing here to offend materialist and atheist readers: Even if consciousness could have a quantum explanation, and even if this role might open the door to the validation of *some* paranormal claims, it certainly doesn't mean there is a God. The two are not necessarily linked. And yes, even if we all already own the New Age holy grail of quantum consciousness, we could all still be no more significant or eternal than meat.

All this uncertainty might be cold comfort, particularly to the materialist tribe that thinks of *might* and *could* and what's *possible* as the sketchy province of believers. But cold comfort, it seems, is all we have—skeptics and believers alike. Up close, in fact, all Hameroff has is cold comfort for those who look to him for a scientifically plausible vision of an afterlife.

IIIIIIIIIII

IN THE YEARS SINCE Penrose and Hameroff first foisted their theory of consciousness on an unsuspecting world, Roger Penrose has in essence absented himself from the ongoing debate. Though Hameroff thought it unlikely, Penrose did respond to my emails and said the following: yes, he has moved on to other things, but he has kept an eye on his Orch-OR theory. And he thinks Hameroff has done an excellent job of defending it. Penrose himself is almost eighty, but he says that, in a couple of years, he plans on diving back into quantum consciousness head first.

He also said he finds some of Hameroff's paranormal musings a bit . . . unhelpful. This shouldn't be surprising. Penrose is a committed humanist, though the British Humanist Association notes, with evident disappointment, that he "is more sympathetic to mystery and uncertainty than some atheists and humanists." (Reading this, one wonders if the British Humanist Association understands that mystery does not require our sympathy, but it does accept acknowledgment.) What might be more surprising is that Hameroff isn't a believer either. "I don't follow any organized religion," he says. "I find some of the Buddhist concepts appealing, but I'm not a Buddhist, either."

In *What the Bleep?*, Hameroff comes off as so happy and chatty, so willing to engage in the more out-there ideas associated with his theory, that I expected a weekend with him to include some incense-burning. But Hameroff is himself beset by people engaged in their own autodidactic vision quests. While I was with him, at the University of Arizona hospital, we went upstairs to check on his mail. There were a few envelopes stuffed with pages of hand-scrawled writings, equations, and predictions about the nature of the world. He even received a fax that looked like a page torn from a mad physicist's graphic novel. The writer, or artist, broke the page up into panels like a comic book, with equations in some squares and crystalline drawings in others, including one of a solitary figure—looking out from its cube into what I guessed was a holographic universe.

As a journalist, I'm trained to look for metaphors. So I took one look at that picture and thought of Hameroff's own recurring childhood dream of standing at the universe's edge. He looked at this image and saw only . . . garbage. He gets so much of this stuff, in fact, that he can't do anything but throw it away. "I feel bad about it," he says. "In a lot of cases, you can just tell, this is their life's work and they're looking for somebody to help them. But if I started looking, I'd never get back to my own stuff."

Hameroff is too busy doing his stuff to read theirs. He likes to spend time outside, hiking. He loves college basketball. He is, in fact, so far from New Agey that when I asked him if he meditates, he looked at me like I'd grown a second head. "Well, sometimes I get quiet and just kind of slow down and gather my thoughts," he said.

Up close, Hameroff is a worker. And his mentality might best be described as pugilistic. His performance at the Beyond Belief conference was typical Hameroff. When I asked him why he gave the speech he did, he said, "They pissed me off. They were basically saying, you know, philosophers and scientists should be the ones making all the decisions. Let's replace organized religion with . . . *us*.' And I just thought, 'This is bullshit.' And besides, everything I said, *is* possible. They just don't want to hear it."

Perhaps the best place to find the real Stuart Hameroff, the worker and the fighter, is at his semi regular conference on consciousness at the

University of Arizona. Whatever Hameroff's critics think of him, they acknowledge the contribution he makes through these conferences, which were first put together by David Chalmers, an Australian philosopher most famous for setting forth the "hard problem" of consciousness at the first Tucson conference in 1994.

In Chalmers's estimation, which has swept the field, it is easy to put down seemingly automatic tasks, like driving a car, to computation no different than an insect darting this way and that. The hard problem is the one of conscious experience: *Why is it I'm aware of myself and my own personal narrative and where this particular drive to the supermarket fits into all that? Why is it that I experience the color red on the stoplight as I do?* "From an evolutionary perspective," Chalmers told me, just recently, "it's hard to understand why we would *need* consciousness."

Chalmers himself is an atheist and *some*thing of a materialist. He doesn't buy his friend Hameroff's idea of quantum consciousness operating inside the microtubules. But he does like the Penrose-Hameroff idea that consciousness is a fundamental property. Consciousness . . . just is. "It's been close to twenty years since I first wrote about the hard problem," he says, "and we're not any closer to an answer now than when I wrote it."

This may come as something of a surprise to the reader, particularly if you're out there surfing the edge of the consumer science magazines, where every new neurological discovery wins headlines. Something that is rarely written about, in the midst of all the hyperbole, is that research at the level of the neuron isn't getting us anywhere in terms of answering the larger philosophical question of consciousness. One physicist I corresponded with, in researching this book, even said he finds all the intense focus on the neuron a bit . . . suspicious. Marshall Stoneham, at the Centre for Materials Research in the Department of Physics and Astronomy at University College London, told me in an email, "I often get uptight after hearing scientists reading too much into matches of, say, thinking about X and the measurable brain signals that can be observable at the same time. It's the Tolstoy problem: *War and Peace,* Epilogue 2. The Russian peasant, seeing a steam engine, will assume it is being pushed by the smoke, whereas the smoke and the motion have common origin and are not cause and effect. And I am cautious about making complex quantum explanations, even though we do believe quantum physics underlies the natural world and how things happen."

Stoneham's take strikes me, and more importantly Chalmers, as being eminently reasonable. We're still working on the issue, after all, and pretending we've got answers before we really do won't serve anyone. The prospect of being the one to nail down the answer, however, is so appealing, that the Tucson conferences are legendary for offering combatants a chance to test their mettle. I spoke to one French neuroscientist, Arnaud Delorme, who said the conferences have become more combative now that Hameroff is in charge of organizing them.

"When Chalmers was in charge, the conferences had a different feeling," he says. "Chalmers is a philosopher, and philosophers admit they don't *know* the answer. They are trying out different ideas. With Hameroff, he is a scientist and so he is used to debates where someone should win and someone should lose because they have the facts."

The field of consciousness studies, however, just doesn't cross that threshold yet. And for a moment I think of Anton Zeilinger, a great scientist who realizes his own inability to reach some definitive scientific conclusion about the nature of reality. I admire Zeilinger because instead of making great assumptive leaps to arrive at some ultimate vision, he seems to have accepted the mystery and sought out the kind of professional who can understand his plight: a philosopher.

In contrast, Hameroff probably does lack the philosopher's light touch. In one of his more infamous exchanges, he battled one of the world's most famous philosophers—Daniel Dennett. In his 1991 book, *Consciousness Explained,* Dennett argued that the problem of consciousness is in fact no problem at all. Our sense of self, the sensation of eating an apple, all these things we call consciousness, are just illusions triggered by the mechanistic computations of different collections of neurons. Dennett's vision of human experience is, in fact, so reductionistic he seems to equate us with zombies—machines that, in essence, only think we're thinking. At one conference Hameroff told Dennett, publicly, "You know, Dan, maybe the reason you like this idea is because you're a zombie. And maybe the reason I see things differently is because, I'm not."

Hameroff told me he was half-joking. But Dennett took offense. "I wound up apologizing," says Hameroff. "I guess he only likes the idea of being a zombie if we're all zombies."

The scientists Nancy Woolf and Jack Tuszynski, who have each collaborated with Hameroff on papers furthering the investigation of quantum consciousness, both say Hameroff sometimes leaves a public talk he's given and turns to them, a bit too late, for advice. "Hey," he says, "did I go too far in there?"

But in person, one on one with some time to reflect, he is far less likely to offend—or step into the ether of his ideas. In fact, when I ask him about the window quantum consciousness opens upon a potential life after death, he closes it. Maybe halfway. "Well, I've certainly never said it's a definite," he says. "But it is possible. These near-death experiences could be the beginnings of that experience, some kind of dreamlike experience of pure consciousness, which is temporary."

"Whoa," I said. "I've never heard you say that before."

We were sitting, at the time, in Hameroff's kitchen. Over his shoulder I could see fat cactuses and an endless sky. "You've never heard me say *what* before?"

"That life after death could itself be *temporary,*" I said.

My head was spinning a bit with the irony of it—death might yet await us, after life after death. "Oh yeah," said Hameroff. "It could be."

He went into a brief explanation. A coherent sense of self in the afterlife might be dependent upon quantum entanglement, but entangled particles can in fact be separated. (There is a potential out here, for believers. Time behaves so differently at the smallest levels of the quantum scale. Maybe once we're all a part of the great cosmic hoo ha, we never experience our own disentanglement—even if it happens.) I let this cold comfort wash over me. In myriad books and articles, mystics prattle on about the quantum afterlife, as if its eternal nature is a certainty. And Hameroff has been quoted on the subject numerous times. But until I asked him, I never knew he saw life after death as a potentially terminal condition.

What so impressed me about this moment, however, was Hameroff's ease, his utter lack of concern. Paranormal belief is so often put down to a simple fear of death. But Hameroff had mentioned this possible, final death with a stunning casualness. I saw and sensed no more emotion in him than I did the previous night, at dinner, when he pronounced his broiled fish mediocre. I laughed, there in his kitchen, about how wrongheaded so much of our thinking is about the paranormal: if he had proposed his ideas because he was a pie-eyed believer, he would hold some kind of spiritual belief. If he had opened a window to the mystical because he was scared, he wouldn't close it so casually.

In the end, I think, he proposes his ideas for one reason only: he found himself, finally, out at the edge of the universe—the man taking over for the boy in his childhood dream. And once he got a chance to take in the view, all he saw there is what's possible: a picture in which the ultimate reality of every line and shape is uncertain; a picture in which our own minds are perhaps more fundamental than even our bodies; in which mind was there at the beginning, like time and space, waiting for us, and abiding after we're gone.

Blazing Saddles

UFOs and the Strange Lights Over a Texas Town

I'm trying to listen to the leaves speak, trying to steal secrets from fishes in the creek.
—Jewel, "Stephenville, TX"

Lee Roy Gaitan didn't know quite what to make of this. Ricky Sorrells was a big, quiet man, and central Texas cool—meaning he kept to himself, he took care of his family, and he never caused trouble. But on this day he came to Gaitan's front door, eyes wet with the threat of tears, telling Gaitan that he didn't know what to do anymore. "I think I saw something," Sorrells told him, "that I shouldn't have seen."

Gaitan could only nod. Because whatever it was that Sorrells saw, Gaitan figured he saw it, too. Weeks later and from a greater distance, but probably the same thing. Lots of people in the trio of little farming communities, Selden, Dublin, and Stephenville, had seen it. But Ricky's encounter was the longest and the closest, and after he was discovered by the media, the most controversial.

"I didn't know what to tell him," says Gaitan.

Well, who would?

The military won't leave me alone, Ricky told him. *They keep flying over my house.* Gaitan was an officer of the law, a police constable. And he stood there and heard Ricky out. Sorrells fought the tears back. He held it together. He thanked Gaitan. Then he packed his big body inside his truck, and he left.

A tiny, dairy farming community of roughly 17,000 people, Stephenville is one of the many small towns in Texas where everyone knows each other and life is largely about routine: breeding cattle, birthing new calves, maintaining water misters to keep the animals cool in summer. Football season is king. The Yellow Jackets, the local high school team, draw 7,000 fans with homemade noisemakers into the stadium to watch them play. But in January 2008, all of that seemed threatened. Literally overnight, the town had a choice—to remain a proud cow town, or to embrace perhaps the most notable mass-UFO sighting in American history.

The attention Stephenville received in the wake of that sighting knocked residents just about flat. An international volunteer group that investigates sightings, the Mutual UFO Network (MUFON), arrived within days to interview witnesses. The streets were filled with television trucks, including a CNN crew, and reporters flew in from as far away as Japan. Angelia Joiner, the only full-time reporter on the staff of the local newspaper, the tiny *Empire-Tribune*, started doing radio interviews and even CNN's *Larry King Live*. Not everyone liked this new status. In fact, when the news first broke, and the *Empire-Tribune* ran a headline about the UFO on page 1, the paper's managing editor cried. She figured the people of her town would become nationally known as fools. And who could blame her?

This country has a UFO problem, after all. You might not have been aware we have one, or thought about it in these terms, but we do have a UFO problem: namely, we don't seem to understand what UFO really means. So here it is: a UFO is an unidentified flying object. So any time we see some object flying in the sky that we can't positively identify, we've seen a UFO. But in the same way the words *paranormal* and *supernatural* have been conflated, we now equate UFO with alien spacecraft.

How this came to be is easily understandable. If we've learned one thing in this book already, people don't like the unknown very much. And so, if we believe we're being visited by other civilizations, we read the piles of books and articles on unexplained lights in the sky, then fill in the massive gaps—with wild tales of alien races, interstellar technology, and government conspiracies. If we don't believe, we hear someone saw an unexplained light in the sky and assume, first, that he's claiming to have seen E.T. Then we figure what he *really* saw was an airplane, Venus, swamp gas, or a helicopter, and he must be a bit foolish—maybe even a UFO nut. Then we laugh.

Our UFO problem is profound enough that it leads some of us to pretend that we *do* know. Our UFO problem is stressful enough that it could even lead a man like Ricky Sorrells to the constable's porch, fighting back tears—and thrust the whole town of Stephenville into the civic equivalent of an identity crisis. And if that's the case, well—just what can we do about our UFO problem, when we seem so fundamentally incapable of letting the unknown be simply that?

STEVE ALLEN THOUGHT HE was headed for a quiet night.

A businessman with his own trucking company, Allen knocked off work a little late that day and drove to a friend's house. The date was January 8, 2008. The temperature felt crisp at 50 degrees. A small fire burned in a ring on his friend's property, and the evening promised to be a pleasant one. The sky in Selden, Texas, bordering Stephenville, is vast, and the elevation where Allen stood is high. But that night the cool temperatures sucked all the moisture out of the air, and Allen, who also flies small planes as a private pilot, marveled at the visibility. Central Texas is not Key West, Florida—it's not built for tourists, but the sunsets are equally spectacular, lighting the sky in soft orange hues. Allen had missed that night's show. But the sky that stretched out all around him was easing into a smooth, majestic twilight.

Allen planned to stay a while, with a few friends, and reached into his jacket for a cigar, which he accidentally dropped in the grass. As he bent down to pick the cigar back up, it happened. Something in the air distracted him. A white light in the sky suddenly blazed into his peripheral vision. As he stood back upright, he saw, moving from east to west, a group of bright white lights. The lights spanned what he took to be a single object from corner to corner. He had never seen anything like this. It was huge, and even more incredibly for an object flying so near and fast, it made no sound at all.

Someone asked the obvious: "Do you see that?"

The rest murmured their assent and fell quiet.

The lights moved across his entire field of view, quick as mercury, until finally they were in a position due west of him. Then they stopped. Allen could never make out any actual craft. All he saw were the lights he took to represent the edges of one object, which now appeared to be hovering. Considering the direction the object had traveled, he figured he was looking at its back end—the lights of which suddenly blinked into various configurations before going suddenly, blazingly white, like the arc of a welder's flame.

Then it was gone.

The thing took off with the easy speed of an exhalation.

The silence stretched on for just a moment. And finally, the three witnesses looked at each other. One of them, who doesn't want his name published, headed for his truck.

"Where are you going?" Allen called to him.

But by the time the words left his mouth, Allen could see the guy had already departed for, well—the Land of Get Up and Go. For him, this unknown light had presented itself as something to be feared. And as his truck rumbled off, down a small dirt road, Allen and his remaining friend, the property owner, Mike Odom, went inside.

You won't believe what we just saw, they said to Odom's wife, or something like that. And they were right. She didn't.

As a kind of reality check, Claudette Odom called the friend who had just bugged out, to see if he could confirm this crazy story.

He answered his cell. "Go outside," he told her.

He was on the main road now, and he could see the lights doubling back. "It's headed your way."

The three of them all bolted outdoors, and sure enough, there it was. This time it came in lower, maybe 1,000 feet above the ground. Still silent. But after just a few seconds, they heard the roar of two military jets trailing it. Selden, Dublin, and Stephenville all sit within 15 or 20 miles of the Brownwood Military Operations Area (MOA). Consequently, the locals see jets conducting exercises often enough to recognize them. These jets were also low to the ground, seemingly chasing the object, and their afterburners were turned on. The noise they made was deafening—a sound out of wartime.

This repeat sighting lasted just seconds. And one of the things that stayed with Allen is that those jets fell further and further behind the object, till they all flew out of sight.

There were other witnesses who saw strange things in the central Texas skies that same evening. In fact, between about 6:15 P.M. and 7:30 P.M., dozens of witnesses saw *some*thing. Reading over the reports later generated in the media and by investigators who arrived at the scene, witnesses sought prosaic explanations. But the object they saw didn't descend, dim, or drift with the wind like a flare. Sometimes it appeared stationary, just hovering, before hurtling off at a rate of acceleration beyond that of a jet or helicopter. And whatever it did, it remained silent.

A police chief in Gorman, Texas, driving along the highway, said he saw one white light at first, which he assumed was a flare, until he noticed that it didn't appear to be descending or dimming. Then it just shut off, like a lamp light with the power cut. And in that same moment, three lights appeared in the space around it. He watched for a while but couldn't match it up with anything he had seen before.

Not far away, Lee Roy Gaitan, the county constable, had his sighting well after sunset. He and his family were going to watch a movie on Pay-Per-View that night, so he went out to his car to retrieve his wallet. The stars were fully visible in the sky. Just above the tree line, he saw a single, glowing reddish-orange light. The trees were a little more than a football field away, and the light seemed to hover just a bit higher than them. The light blinked off, then on again, a little nearer. He went inside to get his family. His wife just snickered and refused. His young son came back out with him, and now he saw numerous white lights that flickered in varying patterns before moving off, in one unified formation, at a terrific rate of speed. Though the lights appeared nearby, and he imagined their rapid departure must have taken a lot of power, there had been no sound.

I spoke directly to a handful of witnesses of what soon became known as the "Stephenville Lights." I spoke to the investigators who went to the scene, and I reviewed the reports they filed and the news accounts that ultimately appeared all over the country. I interviewed Angelia Joiner, the local reporter who talked to more witnesses, more intimately than anyone else could hope to, because they trusted her. And looking over all the material gathered, I must admit, there is no real, clear narrative possible here—or at least no clear narrative that utilizes all the available, credible information and arrives at a neat conclusion. As a reporter, I've encountered the same thing in homicide trials. Even seemingly allied witnesses, there for the prosecution, say things that don't match up. So it isn't just bias at work. It is the confabulation inherent in memory. Did the gunman come in from the east side of the street, or the west? Was the gun black or silver? Big or small? Did the gunman fire one time or three? Did he say anything? I've heard eyewitnesses disagree about all these details. The sighting of a UFO makes the whole process even harder because, well, c'mon: bright white lights in the shape of a massive, half-mile-long rectangle? That wink on and off, assume different patterns and blaze away at speeds that make F-16s look slow? Such occurrences don't fit into our daily store of experiences. And so the mind is set to racing through potential prosaic explanations—and more extreme possibilities. Worse, when a vacuum of data surrounds a mysterious sighting, people at the fringes of belief and unbelief come to dominate the conversation. And the effect is such that witnesses have to think long and hard before talking at all.

Allen, the private pilot, went home that night and told his wife what he saw. He suggested he might call the newspaper. She said *Please don't*. In fact, she told him he shouldn't mention what he saw to anyone. It wasn't that she didn't believe him. She was just being protective. It is hard to come off as sane when describing an event that sounds *in*sane. Allen had seen something he could not identify; and culturally, in America, that means one thing: E.T. He was now the bearer of the stigma associated with the paranormal.

Allen understood his wife's concerns.

He even shared them.

He told her he'd sleep on it. Then, the next day, he picked up the phone and called Angelia Joiner at the *Empire-Tribune*.

IIIIIIIIIII

LIFE HAPPENED FAST AT that point, not so much in Stephenville as *to* it. Joiner's story, published on January 10, with Allen as her main source, brought forward numerous witnesses. The majority of them, including police officers and a judge, have never been publicly identified. Such is the stigma associated with *unidentified* flying objects. But Ricky Sorrells did go on the record.

Sorrells is a machinist, husband, and father, and a few weeks before the January 8 sightings took place in Stephenville, he witnessed an event he later thought was connected. Sorrells was off work that day and decided to spend a few hours deer hunting. He has a secluded property, with neighbors he would need to hike or drive to reach, the nearest one a half-mile away. He got his rifle and stalked into the woods, thinking he'd be alone the entire time. After a few minutes of hiking, however, he got tripped up in some brush and, as he steadied himself, looked up. There it was. Right over his head. Maybe three hundred feet in the air. Barn-metal gray. Massive. He couldn't see from one end to the other, could not see an edge, could not see the sky. His first instinct was to lift his riflescope to his eye and take aim. His breath was coming fast, his heart beating hard. Even as he realized a rifle bullet would do no damage to something this big, he continued looking through his scope. An aircraft loomed above him, without any rivets or visible seams. Its underbelly, however, featured a series of inverted cones, projecting up a very short distance into the object. And slowly, his understanding racing to catch up with his senses, he realized the object made no sound.

Sorrells lowered his rifle, telling himself to calm down, to use this time to study the object, and think, and try and figure out what it might be. He estimates that he stood there, staring up, for a couple of minutes. During that time, the craft seemed to float above him, shifting its position once from his left to his right. Still, he couldn't see an edge. And when the thing departed, there was no disturbance of the surrounding air, no noise to warn him. It went straight up, on a flat plane, so that he couldn't tell where the front was—or the back. And when it lifted, *it left,* out of sight, so fast that he later told me, "If there is a word for that kind of speed, I haven't heard it."

With Sorrells in the mix, this story didn't just have legs—it had lift-off. The Stephenville Lights, which rightly or wrongly now included Sorrells's earlier, daytime sighting, went international. Local hotels booked up. Joiner went from covering school board meetings and the city police blotter to . . . UFOs. The majority of her witnesses didn't want their names published. But they seemed overwhelmingly sincere. They talked to her for the same reason Allen had decided, after his own sightings, to pick up the phone and call her: they were intrigued by what they saw, and they wanted to know what it was. At least three law enforcement officers described separate sightings of a massive airship floating over the town. One even got a lock on the object with his radar gun. And they came forward so people would know the civilian witnesses weren't lying.

Right away, the military issued denials. Multiple witnesses had described jets flying at low altitudes. But Maj. Karl Lewis, a spokesman for the 301st Fighter Wing at the Joint Reserve Base Naval Air Station in Fort Worth, claimed they had no planes in the sky that day at all. He also went on to claim the sightings might be attributable to sunlight glinting off an airliner—a kind of titanic nonexplanation, given both the twilight conditions and the witness accounts.

The Mutual UFO Network, a collection of volunteer investigators, came to town and interviewed people. Visiting media and the MUFON members asked Joiner's advice on everything from where to stay to whom to talk to, to where they might find good barbecue. The MUFON reps quickly made a series of Freedom of Information Act requests, citing the reports from local radar towers as public information. And they ultimately interviewed a couple of hundred people.

In a town as generally quiet as Stephenville, this was a carnival. Sarah Cannady, at a bookshop called the Literary Lion, invited witnesses to write sighting accounts in her store. Barefoot Athletics, a T-shirt shop, usually hawks school spirit athletic wear, with slogans like, "Dribble it . . . pass it . . . we want a basket!" But they knew an opportunity when they saw one, and they started printing and selling UFO-themed shirts as fast as they could make them. Kids walked around town in tinfoil helmets, waving to honking motorists like the central attractions at a great parade. Some made a joke of guarding Moo-La, the fiberglass cow and semi-official town mascot on the courthouse square. Restaurants and shops put up signs advertising where UFOs should park. And the Fiddle Creek Steakhouse started selling what they called an "Alien Secretion" shot: 3/4 shot of Malibu rum, 3/4 shot of melon or Midori liqueur, 1/2 ounce of sweet and sour mix, 1/2 ounce of pineapple juice.

It wasn't hard to see, and certainly the town's political leadership was aware, that circumstances were forcing them to make a choice: To Roswell, or not to Roswell. That similarly small town, in New Mexico, marked the supposed 1947 crash site of an alien craft. And Roswell had come to fully embrace that hotly disputed event as the centerpiece of its national and international identity, catering to year-round UFO-obsessed tourists with a UFO museum and an annual UFO-themed conference and parade.

Stephenville had the opportunity to pull off a similar transformation. In fact, a few days after the story broke, city political leaders and representatives from the Chamber of Commerce even got together and talked about it in an emergency meeting.

From an outsider's perspective, Stephenville is a small town from another time, its reputation staked on dairy cows, rodeo, and its status as one of numerous American hamlets calling itself, "The Cowboy Capital of the World." Nearby Dublin has the original Dr. Pepper soda plant. Tarleton State University gives the area an intellectual center. And the population claims two celebrity residents: Ty Murray, arguably the best competitive bull rider in history, and his wife, the pop singer Jewel. But Stephenville has, shall we say, room to grow. And while no one seemed sure how long the UFO might remain afloat as an economic engine, right now it was soaring. The local T-shirt shop was selling thousands of copies of their new design across the globe. "Stephenville, Texas," it read, "The UFO Capital of the World," atop a picture of a classic gray alien with a Stetson hat. Stephenville got calls from businessmen wanting to help them capitalize on this opportunity. Travel companies wanted to set up regular trips to the sights where witnesses saw unexplained lights. They wanted to talk about a museum and a gift shop. They wanted to bring in tourist revenue and put paying customers in local hotels and restaurants.

A whole new identity.

There for the taking.

But Stephenville said . . . no.

They didn't want any part of UFO tourism. They liked what they are, just fine. But in a sense, such a decision is only mostly up to them. So by the time I got there, a little more than a year after the sightings, I found a place that was perhaps not the UFO or cowboy capital of the world.

I landed in a place where the people clearly wanted to be friendly. When I wasn't talking about UFOs, in fact, they greeted me with big smiles and talked about their town with pride. But when I fessed up, and said, "I'm here about the UFO," the whole character of the conversation changed. Some people talked. Some even told me they saw it. But about half the people I approached smiled tightly and said something along the lines of, "Oh, how nice," then lapsed into taut silence. At a local shopping center, where I stood outside questioning strangers for an entire afternoon, a half-dozen people heard me say, "UFO," and just walked away without responding at all. The mayor of Stephenville started ducking me before I even arrived, failing to return repeated phone calls about my upcoming visit, then shunting me off to the Chamber of Commerce. The city's former mayor, who had presided over the town when the lights appeared, also never responded to emails or phone calls.

Stephenville had made a decision to keep UFOs from taking over the town. And in doing so, they had seemingly created two Stephenvilles: one anxious to know the source of those unexplained lights, and another that was simply anxious.

IIIIIIIIIII

THERE IS REASON TO be wary of this entire topic.

The mind does play tricks.

In researching this story, I ran across Tim Edwards.

A restaurant operator in Salida, Colorado, Edwards had appeared on the now-defunct television show *Sightings*. The program broadcast a video that

Edwards had taken of what he thought was a massive UFO, sitting way up in the sky, with an array of lights that flashed back and forth, from one side of the craft to the other.

The video looked inconclusive, at best, but the incident reframed his entire life. Edwards changed the name of the restaurant he ran from The Patio Pancake to E.T.'s Landing. He credited the event with making him a more spiritual person. And he kept his video camera handy. Over the ensuing years he recorded what he claimed were UFOs pretty much all the time.

More than a decade later, I called Edwards to see if the conversion stuck. His family informed me that he'd died of a heart attack. But he was a believer in alien visitation, they assured me, to the last.

I felt a bit uncomfortable, in light of this news, but asked anyway, about what I called "The Spider Web Theory."

Edwards's family defended him and his sighting to me. But the truth is, their loved one's conversion experience has been convincingly undermined. Dr. Bruce Maccabee, an optical physicist and a longtime supporter of the idea that aliens are visiting Earth, is the one who did the debunking. Maccabee told me he conducted an experiment after Edwards's story aired. He concluded the visual in Edwards's famous tape could be produced by a thin film of moisture on a spider web. At the right angle, in the right light, a single, nearby strand of spider silk, stretched taut across the sky, can look like a spaceship, thousands of feet overhead, the refraction of sunlight on water creating the illusion of strobing lights.

Because human perception can be so easily tricked, skeptics do some of their best work with UFOs. I urge all readers to take a look at the Rendlesham Forest case. It is long, colorful, and complicated. But the gist of it is that a UFO drifted about the forest surrounding a U.S. Air Force base in Britain. At one point, the base commander even lit off into the woods, trying to find the source of the strange light his men reported. He made an audio recording of his hunt, and skeptic Brian Dunning has since synced it, inserting a beep at five-second intervals. Sure enough, every five seconds, during the most dramatic section of the recording, the commander sees the mysterious light again. But as it turns out, five seconds is in fact precisely the time it takes for the nearby Orford Ness Lighthouse to make a complete rotation. The soldiers, Dunning argues, were fooled by light coming through the trees all right—from the Lighthouse.

Of course, skeptics don't always make their cases so impressively. In the wake of the Stephenville sightings, for instance, they proposed sundogs, flares, military jets, or even a rare "superior mirage," even though the sight reported didn't match up with any of those explanations. At one point in our country's tortured UFO history, longtime debunker Philip Klass strenuously argued that some form of ball lightning could explain many UFO reports. Which is really funny. Because one of the more interesting corners of current science is the ongoing debate about whether or not ball lightning even exists. And here's one more for flavor: in the great flap of UFO sightings around Washington, D.C., in 1952, Klass clung to the explanation of a weather inversion, which can produce false radar hits. The problem for Klass was that the reports from D.C. included both radar evidence and multiple eyewitnesses. Worse, believers also trotted out their own equally credited investigators, who pretty convincingly argued that a weather inversion couldn't account for the radar data.

Just admitting that there is a kind of standoff in regard to the evidence, however, doesn't seem permissible for either side. So believers continue to point to D.C., 1952, as an E.T. fly-by. Skeptics call it all explained. And emotions, on that case and others, run hot. There is perhaps more personal enmity in this field than any other within the paranormal. Klass, in fact, left behind a "curse" (long) before his death: "To UFOlogists who publicly criticize me or who even think unkind thoughts about me in private, I do hereby leave and bequeath THE UFO CURSE," it reads. "No matter how long you live, you will never know any more about UFOs than you know today. . . . As you lie on your own deathbed you will be as mystified about UFOs as you are today. And you will remember this curse."

In turn, some in the UFO community cursed Klass's memory: "The world is better off without him," UFOlogist William Moore was quoted in one pro-saucer newsletter. "My sainted grandmother told me not to say anything about the dead unless I could say something good. He's dead. Good!"

These examples are extreme, but indicative. The tenor of the UFO debate is generally embarrassing for everyone, and even the jokes are ugly. The combatants speak in their own language, like members of high school cliques: UFOlogists, as they like to be called, are dubbed "saucertards" practicing "pseudoscience." And the skeptics are accused of being "in on the alien conspiracy" and denounced as "pseudoskeptics"!

But what leaps out at me is that the skeptics and the believers share one startling point of agreement: the vast, vast majority of so-called UFO reports can be convincingly identified as something altogether earthly. Of those that remain, there is either too little data to come to any conclusion, or the data are plentiful and the object remains . . . *unknown*. *The Condon Report*. *Project Blue Book*. MUFON's own data. They all agree on this.

In the *sturm und drang* of debate, however, believers and skeptics just can't seem to let the unknown stay just and only that. And so, in cases where no clear answer exists, they continue the argument, each fighting for their points of view; and when the rest of us, without the same emotional investment in the debate, turn on the television and see a UFO report, we usually become witnesses, too: to the equivalent of Valium talking to mescaline—two seemingly inebriated factions engaged in a long-term war. My favorite illustration of this is when the skeptic James McGaha asked two UFO witnesses, during a *Larry King* appearance, "Are you *qualified* to look at the sky at night?"

I'm guessing he was wondering if either of them was an astronomer. But do we need an accredited university to *qualify* us to use our eyes after the sun sets? I believe there is a more profitable way to approach this topic.

A lot of great thinkers are on the record with the observation that the sudden flash of insight, the gut hunch, and the creative leap are what truly push science, and humanity, forward. Edward de Bono has spent the past twenty-odd years, in fact, traveling to fifty-two different countries and consulting with some of the world's largest corporations, including IBM and DuPont, teaching his concept of lateral thinking. Perhaps the foremost expert on the topic of creative thought, de Bono makes an argument that sheds an awful lot of light on our collective UFO problem.

De Bono argues that the West's tradition of settling disagreement by debate or argument is an example of *over*reliance on logic. In debate, the best debater wins. In argument, the person whose case best fits the rules of logic and the *current* evidence wins. And yet, scientists themselves report that their best discoveries often don't come when they are slumped over a pen and paper, poring over data, thinking oh-so-logically. They come, as neuroscientist Terry Sejnowski put it in chapter 3, when they're in the shower. Or brushing their teeth. Or sleeping. Biochemist Albert Szent-Györgyi said science depends on "seeing what everyone has seen and thinking what nobody has thought."

In sum, logic is a partner to freer, associative thinking. Logic can help define the contours of a problem, but sometimes, in order to find a solution, we have to go beyond such strictures to find the unseen path.

My favorite example of sudden insight is that of the Irish mathematician, William Hamilton, who spent fifteen years on the problem of generalizing the square root of –1. The answer came to him whole and suddenly, as he walked in the street. Lacking a pencil and paper to write the formula down, he took a penknife from his pocket and carved it into the stone of a nearby bridge. "An electric current seemed to close," he later wrote, "and a spark flashed forth."

Dr. David Jones, at Newcastle University, long collected such accounts and gave a series of BBC lectures on the topic of inspiration. He cited the result of a survey of 1,450 American scientists conducted by two research chemists in the 1930s. Their finding, subsequently published in the *Journal of Chemical Education*, was that "hard, rational thought" traced the contours of the problem to be solved, not its answer. Think about the problem, they advised, then go do something else. Don't allow worry or anxiety into the brain. And suddenly, while driving a car or sleeping or bathing, the answer will pop, suddenly, and seemingly fully formed, like the mathematician's formula, into view.

De Bono's methods are designed to help people get outside the rigors of the logical thinking we're usually trained in, to create these flashes of insight and make new conceptual connections on a regular, predictable basis. Logic does have its weaknesses: Logic is based on society's current storehouse of accepted knowledge, and as we've seen, that storehouse changes all the time. To refer back to the previous chapter, five years ago the average physicist would have been happy to take the position in a debate that quantum effects play no important, measurable role in biological systems. The available evidence would have supported that case, while opponents would have been arguing mostly from *possibility*. But today, biologists are chasing after quantum effects in plants and animals. The very grounds of the debate have shifted, and in just a few years' time.

In his book, *Oracle Bones*, Peter Hessler describes how the rulers of China's Shang Dynasty pondered cracks in turtle shells. In the sound of the shells snapping, it was said, they heard the voices of their ancestors offering advice. From our modern perspective, the cracks meant nothing. But through the lens of creative thinking, these cracks gave their interpreters a place to stand outside the rigors of logic, to dream, to allow their minds to associate freely and make new connections. Oracle bones comprised a regular practice of a stable society, a dynasty that lasted hundreds of years.

In this respect, people who look at that relatively small number of unexplained UFO cases and wonder if aliens are visiting the planet, who carry it along with them as a possibility, are far from stupid. They are merely engaged in some creative thinking. And if, some day, they turn out to be right, their interpretation of the reports they've read of unidentified flying objects will have led them to the right conclusion— like cracks in a turtle shell—faster than current evidence alone could have gotten them there.

The possibility of visitation holds more credibility than skeptics, or the rules of formal debate, might allow. Given the size of the universe, our distance from other potentially habitable planets, the energy sources and the top-end speed of which we're aware, it seems exceedingly unlikely another civilization could find us and get here. And personally, I'd like to see incontrovertible evidence before I start believing aliens have visited the planet. But I also don't agree with the skeptics when they *ridicule* such belief. After all, an alien society would be just that—alien. Debates about their methods of propulsion, and their motivations—Why would they bother with a dairy farming community?—are by definition utterly hopeless. Their supposed inability to find us and get here is a product only of our own current understanding—and might hold absolutely no relationship to reality.

Of course, none of this means E.T. exists. And when I attended a weekend "UFO Awareness Day" here in Philadelphia, I was stupefied by the level of emotional commitment some attendees had to their far-out concepts: some spoke of a "hidden" planet, Nibiru, that they believe will impact Earth in 2012; some talked about reptilians, an evil race of lizard-looking aliens with forked tongues; and many harped on alien abductions.

Like the skeptics who argue against them, they take it all so seriously. And in this point of relatedness, I think, the combatants on either side too often fail to acknowledge what the rest of us know about aliens: they're fun.

From rock star David Bowie's androgynous alien creation, Ziggy Stardust, to the identity politics of the recent film *District Nine*, aliens have provided a creative launching pad for explorations of everything from fashion and human sexuality to issues of race and international relations. Aliens in fiction have allowed us a window through which we get to see ourselves, and they serve as an oracle bone for our culture. That said, some people don't have the luxury of considering the question of E.T. solely at the multiplex. Some people have the issue thrust upon them. The people of Stephenville, for instance, never asked to play host to a paranormal controversy. The people of Stephenville just looked up.

IIIIIIIIIII

IN THE WEEKS AND months after the sighting, it slowly became apparent that the Stephenville Lights were good candidates to stay in that unknown category. The military's behavior was one reason. After two weeks of maintaining they had no jets in the area, they reversed course, saying ten F-16s had been flying maneuvers outside the nearby Brownwood Military Operations Area. They proclaimed their initial denial "an internal communications problem."

After spending about a week in Stephenville, I left believing that the military was either directly involved or had some interest in whatever flew over this small Texas town in January 2008. Again and again, locals told me that the days after the sighting were filled with flyovers by military jets and helicopters. "I had a lot of media who were in town covering the story, asking me," says Angelia Joiner, the reporter from the *Empire-Tribune*, " 'Does this happen every day?' "

"I've lived here my whole life," says Matt Copeland, one of the owners of Barefoot Athletics, "and I had never seen anything like it. There were jets flying overhead, sometimes really low, all the time."

The military denied they were doing anything unusual. "I think they were trying to discredit the witnesses to the national media," says Joiner. "Because if there is all this stuff flying over here, then maybe people just got confused."

Believers tend to think of military conspiracies to cover up the existence of aliens. But I wonder if the military was covering something up, whether that something was some secret military craft. If so, this would not be the first time witnesses saw a paradigm-busting technology and thought it extraterrestrial. UFOlogists concede the U2 spy plane and the sleek, black, triangular stealth bomber were likely responsible for waves of UFO sightings. The object in the Stephenville sightings had flown outside the military operating zone, but who knows? Maybe Stephenville, on January 8, 2008, had in fact been the site of a test run for a new military project?

I bring this up because a military craft is one of those more logical possibilities (and one that fits the reported facts better than flares or some odd weather phenomenon). But I also mention it for another reason: if something radically new was flying through the Stephenville area that night, I'd argue for the overall accuracy of the town's eyewitnesses. Certainly, if there was some sort of advanced military technology tooling around, their testimony that they saw something they couldn't explain would seem to me completely correct—and no laughing matter at all.

In this sense, I see the people of Stephenville as standing in for all eyewitnesses to the paranormal—ordinary people who didn't ask for weirdness, or to be seen as somehow strange themselves, yet had this cartoonish mantle thrust upon them. In the early days after the sighting, they weren't out there telling the assembled media they saw craft from another civilization—just that they saw something that outstripped any technology they know about. I'd argue that in this claim, the evidence suggests they were correct. And yet, given our nation's UFO problem, they found their own credibility undermined in subtle ways.

Allen, Sorrells, and Joiner note that the national press chose to report the same culturally coded details, again and again: Sorrells was deer hunting when he had his sighting; Allen claimed the UFO was "as big as a WalMart."

As a journalist, I can tell you: I'd use the WalMart line, too. It's colorful and evocative. It's a great quote. But it is also true that, like Sorrells's hunting, the WalMart reference seems to peg Allen in a specific socioeconomic and cultural place. "I regret saying that now," Allen told me.

At the time the media glare got hottest, Pam Kinsel, a member of the high school science club, spoke to a reporter and gave voice to the community's worst fear: "[The sighting] makes us look like we're a bunch of retarded hicks," she said.

At the time, in early 2008, the depths of this country's UFO problem had recently been revealed in dramatic fashion: Ohio Congressman Dennis Kucinich was running for president of the United States during an October 2007 Democratic primary debate when a UFO shot him down. It was the usually sober-minded political journalist Tim Russert who pushed Kucinich under the equivalent of a paranormal bus: "This is a serious question," he began, lest anyone think he was joking. "The godmother of your daughter, Shirley MacLaine, writes in her new book that you sighted a UFO over her home in Washington State."

The crowd can be heard at this point, laughing and gasping as Russert continues, "[She writes] that you found the encounter extremely moving, that it was a triangular craft, silent and hovering, that you felt a connection to your heart and heard directions in your mind. Now, did you see a UFO?"

"I did," said Kucinich, "and the rest of the account—"

The crowd started groaning and gasping again, and Kucinich reminded the crowd what UFO really means: "It was an *unidentified* flying object," he said. "Okay? It's unidentified. And I saw something. . . ."

The crowd buzzed with laughter, the point made: Woe be unto you, who admits to seeing something you can't identify.

The elfin-looking, pointy-eared Kucinich was a marginal candidate. He was never going to win the primary, anyway. But his run was effectively ended by Russert's question.

He had been wise enough, over the years, never to bring up the sighting himself. But MacLaine had ended his hiding for good, with her book, *Sage-ing While Age-ing*, which described a twenty-five-year-old sighting at her home. Intriguingly, they were near a military base, and military helicopters were spotted trailing after the object there, too.

In the wake of Russert's "serious question," *The Wall Street Journal* published a front-page story portraying Kucinich as silly. And the rest of the media piled on—the twenty-five-year-old sighting he had never discussed now serving as a metaphor for his frivolous candidacy.

This all went down just a couple of months before the people of Stephenville had their sighting and rendered them keenly aware that the Stephenville Lights marked them, in many minds, as an unsophisticated tribe. But the people I found were smarter than all that. They understood how deer hunting and WalMart had been used as emblems of the lower class, to be slapped on their foreheads by a media eager to play into cultural stereotypes. And over the course of my time in Stephenville, I encountered several locals who sized me up on the Internet and told me all about it in a manner that was at once straightforward cow-town Texas and utterly contemporary America.

"I checked you out on Google," Frank Burke, a spirited old man, told me. "I saw the kind of stuff you wrote at your old job . . . the *Philadelphia Weekly*? You're all right."

I'm not going to argue that the people of Stephenville qualify as technophiles just because they learned to use the Internets well enough to work the Google Machine. But I was impressed that they had gone deep into my history, beyond the magazine where I work now to my previous newspaper job, and that they used my clips on crime and politics and traditional journalistic subjects, to assure *them*selves, the UFO witnesses, that *I* wasn't some kind of nut. And I note that if these are modern-day hicks, then hicks as we have ridiculed them simply do not exist in great numbers anymore.

The majority of the media, however, seems to have missed this particular memo.

After reviewing all the news accounts I could get my hands on, I've determined that the most telling, wrong-headed cheap shot to the people of

Stephenville was ultimately delivered by a writer from the *Los Angeles Times*: "The night sky above Stephenville is a jet black canvas that seems the perfect backdrop for the sharp white specks of stars," the *Times*'s subsequent article reads, "and any imaginings of strange glowing lights."

The implication that the witnesses could have deluded themselves by looking into an average Texas night sky is clear. But an absence of ground lighting, allowing the stars to appear in greater relief, would only seem exotic to a city dweller. Like, say, a writer from the *Los Angeles Times*. In central Texas, these vaulted skies, these jet black canvases, are simply called "night." And as it turns out, there is good reason to think that on January 8, 2008, the people of Stephenville really did see something in the skies over their town.

The radar data ultimately published by MUFON, in fact, suggest one object traveled at varying speeds across the region for about an hour, hovering or moving at less than 60 mph most of the time, but occasionally accelerating to speeds more than 500 miles an hour. Whatever it was, it flew without a transponder—meaning it wasn't official air traffic, and if it *was* military, it didn't particularly want to be tracked. The data in the MUFON report also included another pair of radar hits corresponding to the testimony of several ground witnesses, including Allen, that suggested an object was flying at about 2,000 miles per hour.

Of course, this too has been explained away by skeptic James McGaha, who told *Popular Mechanics* he thought some kind of "radar scatter" was responsible for the entire data trail MUFON found. "They had a huge amount of data," McGaha said, "and they just pulled a few bits of information out of it and drew a line."

I interviewed Robert Powell and Glen Schulze, the men responsible for the MUFON report. They point out a couple of things to undermine McGaha's criticisms: one, multiple eyewitnesses corroborate the data they got from the radar reports. Further, the kind of "scatter" McGaha was talking about reveals itself: "With scatter," says Schulze, a retired radar operator, "you'll see the radar makes its sweep, in a circle, and when it gets a false ping there are multiple supposed returns, all around each other. The data you're looking at tells you these are false hits. This wasn't like that at all. This was one clear path, traveling over an hour, in the same direction, at a consistent speed, with one hit per sweep. This was like a clear track in the snow."

Further, Schulze and Powell took a close look at the data yielded by the towers that produced those key hits, looking for standard signs of malfunction in the hours before and after the sighting. They found none.

Again, this doesn't mean the source of the Stephenville Lights was extraterrestrial. But it does suggest that the witnesses saw *some*thing real and solid enough to yield consistent radar returns. And whatever it was, the fallout was real enough for the people of Stephenville.

Joiner quit her job at the newspaper. The marching orders coming out of the city's meeting to reject a new, UFO-based identity included a call for the newspaper to back off the story. The *Empire-Tribune*'s publisher is said to have attended the meeting. And really, on a journalistic basis, I can't blame the newspaper brass for tacking back toward normalcy. Small-town newspapers stay relevant by covering the *small* stuff, not by investigating military cover-ups or interstellar travel. Joiner, however, felt returning to her beats would leave the witnesses twisting uncomfortably in the wind. "They still wanted answers," says Joiner. "And they had come forward because they trusted me. So I wanted to get those answers, too."

She left the paper and started running a UFO-related web site and podcast. She is also pursuing a book-length treatment of the Stephenville Lights.

Gaitan paid a price of sorts, too, at least for a time. The county constable spent many long nights driving all over the countryside, trying to catch a glimpse of the craft again before he realized he needed to get ahold of himself. "It was just taking up too much of my time," he told me. "I had to shut it down."

Though his Facebook page lists *UFO Magazine* among the products he recommends, Gaitan is no longer driving around, looking for weird lights.

The fallout includes the whole town, to some degree. Mistrust lingers between local residents and the Air Force personnel in charge of the Brownwood Military Operations Area. Mark Murphy, a city councilman at the time, says he had it out with personnel at the Air Force base. "It was, like, two weeks, we were seeing jets and helicopters over the town, day and night. I called them. They said they weren't doing anything out of the ordinary. That's crap."

In Murphy's estimation, the whole episode was some exercise conducted by the military. "I certainly have no proof it was extraterrestrial," he said.

And then there was Ricky Sorrells.

Joiner proved instrumental in getting me together with Sorrells, who doesn't plan on giving any more interviews. The initial wave of publicity was difficult. After he drew all that attention to himself, he says, military copters and F-16s flew over his property at odd hours. Then he got a phone call from a man identifying himself as a lieutenant colonel who wanted to meet with him. When Sorrells hesitated to agree, the man reacted angrily, insisting on a meeting. Sorrells advised the man not to cross his perimeter fence. "If you are who you say you are," said Sorrells, "why don't you quit flying in the airspace over my property?"

"It's my airspace," the man replied. "But I'll quit flying my helicopters overhead, if you quit talking about what you saw."

Sorrells didn't commit to keeping his mouth shut. And he never did meet with the lieutenant colonel. But not long after that conversation, he woke up to the sound of his dogs barking.

It was around 1:00 A.M.

He went to the window and saw a man standing on his property, staring at his house through a steady rain. Sorrels grabbed a rifle he keeps in his bedroom and, without waking his family, went to the backdoor. The man stood forty or fifty feet away, between Sorrels's car and pickup truck. The man wore a heavy parka. But Sorrells couldn't make out any detail in his face or clothing that would have specifically identified him as a member of the military. The man's face seemed to be angled directly toward the window Sorrells was looking out from, however, and Sorrells felt sure the man had spotted him.

He wanted to open his back door but felt like he'd be at a disadvantage. The man across from him had plenty of cover. Ricky would have been an easy target, if that's what this was about. The pair stood in those same positions for many long seconds. The man rocked back and forth slightly on his heels in the rain, biding his time. Then he slowly turned and departed.

In the morning, Sorrells went outside to look around. He found the man's footprints. And he also found something else: a bullet, right where the man had stood. The bullet was just sitting there on the open ground, flecked with mud from the rain. Sorrells took the visit as a further warning.

I met Sorrells, Joiner, and her husband in a Mexican restaurant and had a couple of beers. Today, Sorrells isn't the UFO witness we usually see in media caricatures. He isn't waiting for the space brothers to make life on planet Earth better, or the reptilians to land and destroy us all. He is in fact a reluctant, regretful witness. "I wish I'd never seen it," he told me. "Whatever it was, I wish I never did look up."

Which is, of course, how he ended up at Gaitan's door.

"It got pretty bad for me for a while," says Sorrells. "I didn't really know what to do."

During this period of crisis, however, a man made contact with both Sorrells and Joiner. The man coached them on how to deal with these strange events and became someone Sorrells could feel comfortable talking to about his sighting. Neither Joiner nor Sorrels would reveal anything about this mystery man, who wants to keep his own identity a secret. And as a result, I can make no guesses as to whether he is retired or active military, a cool, old, beer-drinking dude, or an imp from the Seventh Dimension. But Joiner's husband, Randell, showed me some of what they've learned, if not from this suspicious mister, then from the whole experience—about how to maintain a sense of humor in the face of uncertainty and how to take the cultural expectations of outsiders and use them like a shield. "You know, we aren't the kinds of people who would ever think about things like UFOs," he told me. "This is Texas. We think about things like football, barbecue . . . and steers!"

He smiled at me the whole time, out from under the brim of his cowboy hat, clearly playing with the way his town had been represented. Then he winked. And we ordered another round.

IIIIIIIIIII

IT IS FRIDAY NIGHT, and the lights are on at the Yellow Jackets stadium, the church of high school football filling all the pews.

This is part of the life the town's leadership wanted to return to in rejecting the opportunity to become a UFO mecca. "It is a branding issue," said July Danley from the Chamber of Commerce. "And there was a lot of talk in the community about maybe capitalizing on the sightings in some way. But it was decided that we like who we are. We like being the cowboy capital. We like what we have to offer, and we didn't want to change that."

I learn a lot about Stephenville, and its aspirations, peering through the lens of the Yellow Jackets football game. The cheerleaders engage in some old-fashioned, *get that ball over the line*-style chants. And the crowd, five thousand strong tonight, shakes ball bearings inside oil and coffee cans whenever they want to make some noise—a local tradition. Timeouts are filled with folksy ads from the stadium announcer, urging people to visit some of the local barbecue joints and steak houses. And whenever any player stays down on the ground with an injury, players on both teams take one knee until the fallen player is brought to his feet.

I feel as if I'm looking back in time, to a version of life more closely associated with the 1950s than the new millennium.

I maintain a relatively low profile through the course of the game. There is a friendly old couple next to me, however, and as I tell them what brought me to town, I notice a mom a couple of rows back wrinkle her nose. She quickly leads her kids to another area to sit. No UFO talk for them.

The old couple just laugh, a bit darkly. "The UFO's not our favorite thing," the old woman, Linda, says. "We know some of those people, and I believe they saw something unusual, but . . ."

Her voice trails off there. She clearly doesn't want to say too much. Beside her, her husband smirks the whole time.

"You don't believe it?" I ask.

"I didn't say that," he said. "It's just . . ."

He breaks off and makes a fluttering motion with his hands: "Who knows?" he says.

"It's a little silly," suggests Linda. "The attention. No one knows what they saw. I tend to think it was military."

Her husband nods more emphatically at this than ever.

"They didn't come out with the truth at first, about the jets," she adds, "and believe me, people around here, we know jets. I think whatever it was, they were involved somehow and maybe they were into some experiment. But those people, they didn't make it up. I believe they saw something."

In her cold appraisal of the military's behavior, I think, she alighted on one of the most important aspects of the Stephenville sighting. But it is in her regard for the witnesses that I believe we can all most readily learn something.

The truth is, we don't have to treat the so-called paranormal the way we do. We don't need to bathe in it with the believers, or strenuously deny its existence, like the skeptics. And we don't have to turn the whole thing into a fight. The people of Stephenville seem to have struck up a bargain among themselves, in which the believers go on believing, and the skeptics go on being skeptical. Either way, on Friday night, just the same, they all go watch the Yellow Jackets play football.

I talked to dozens of people in Stephenville, if not a hundred—random people in stores and on the street. Some decided they would rather not talk about the UFO. But when they did talk, when they *did* offer me their opinions, I never heard a whisper of judgment creep into their voices. I never heard anyone called a name, never heard anyone else's point of view dismissed outright.

Was that some show of solidarity they put on for the reporter from Philadelphia? Maybe. In some instances, yeah. Might have been. But if that's the case, let the show go on.

Stephenville has made a decision, publicly, to let the unknown stay just that. And they have further chosen not to let their own collective identity be shaped by an enigma. This is not to say that life goes on here as it always has. The people here are warier than before. The mayor never did return my phone calls. Some people wouldn't speak to me at all. The Stephenville Lights did make Stephenville, Texas, a *different* town—a town that has felt the heat of the nation's spotlight.

But that's okay.

A UFO flew overhead, and ever since, the people here have been doing the best they can to rise above it.

Was There a Ghost in My House?

The Unexplained Noises That Fueled a Childhood Mystery

It is wonderful that five thousand years have now elapsed since the creation of the world, and still it is undecided whether or not there has ever been an instance of the spirit of any person appearing after death. All argument is against it; but all belief is for it.

—Samuel Johnson

I was about six years old when the trouble started, and I retain only fragmentary memories of what happened: I remember the sound, big and booming. To me, the noise suggested something was angry—and trying to break in through the roof. Once, I can remember, the sound woke me up out of bed. I shared a room with a big brother, Dave. He had already stood up and turned on the light. Our sisters were down the hall. We could hear them, hollering to us. But the booms seemed to come from the roof over our heads.

"What's happening?" I asked.

In response, Dave stared up at the ceiling and put his hands over his head—the way I'd found stressed adults sometimes did to keep their heads from, you know, popping off. "I don't know," he said.

His voice sounded weak, enervated from his own sense of not knowing. The banging stopped for several long seconds. Then a particularly loud thump sounded, and for a moment I held my breath. I don't remember much else, except that eventually we all gathered out in the hall, like tenants turned out by flooding in our apartments.

My parents came up with a cover story for me, the youngest. And now I can see the ways they tried to give that story legitimacy. I had just got home from school, and my mother was in the front yard, talking to a neighbor, who knew of our trouble. "How's it going?" she asked. "Is it still happening?"

"It's still happening," my mother responded, then looked at me: "Right honey?"

She smiled, still turned toward me: "There are raccoons that jump up and down on the roof!"

I went and found a hole in the ground. I called my mother and our neighbor to look. "It's a raccoon's footprint," I said.

My sisters claimed worse things than banging. One suffered frequent, awful nightmares. Both claimed the covers were pulled from them at night—their sheets and blankets seemingly clutched away in unison, to fall on the floor in a heap at the foot of their beds; an old woman appeared in their bedroom and walked right through the door.

The banging started in 1975 and lasted for nine months, maybe a year, emanating from the roof and walls. Morning on the last day we ever heard it started strangely. My mother put on a dress and makeup on a Saturday morning, like she was going out for dinner. Then she explained that our parish priest was coming over, to bless the house. "We never got the house blessed when we moved in," she said, lying to me about what was really going on. "So we're having him do it now."

Father Crowley showed up maybe a half-hour later, sprinkling holy water in the corners and praying in Latin. I remember that vividly, but I recall only a few details of what happened that night. Mostly, I just remember being scared. And the thumping going on above me. I'll get back to that. But first, a little context is in order.

Looking back, we had what believers would refer to as a "classic poltergeist experience"—a spirit making itself known by either moving objects or making noise. Skeptics would likely hear this story and think our "trouble" emanated from malfunctioning water pipes and overactive imaginations. The impasse between these two conflicting views is such that, since college, I have told this story only rarely. "The Family Ghost," as I call it, is a subject probably best kept to myself. I'm a reporter, after all, and my own professional credibility hinges on *my* reliability as witness. I have not a doubt in my mind that there might be some potential employers, down the line, who won't like this book, even if they never actually read it.

So, you might ask, *Why do this?*

Of course, I asked myself that same question. And my response starts with something fundamental: I'm going to write about this, the Family Ghost, because these stories are with us, whether we like them or not.

IIIIIIIIIII

SINCE THE BEGINNING OF recorded history, men and women have told tales of ghosts and hauntings and things, in our case, that went *boom* in the night. These events are, I'd argue, a part of who we are—if not as individuals, then certainly as a species.

Besides that, as a journalist, I adhere to a code of ethics that requires me to explain any personal connections I might have to the subject at hand. It just doesn't seem right for me to write a book about the paranormal without letting you know where I come from. Most of the events and places that colored my childhood have fallen away. My family stopped attending any regular organized religious services when I was twelve. I've moved more times than I can count. But I'm resurrecting the Family Ghost, so to speak, because as a society, we hear plenty of ghost stories from people who believe everything they hear, and from people who don't think anything labeled "paranormal" could be possible. I think it's time to hear one of these stories from someone who, no matter what he believes, is prepared to focus on what he knows. And folks, that ain't much.

Consider my own memories: every last one of them could be false. In one of the most famous studies demonstrating the unreliability of memory, 120 people who attended the Disneyland theme park were shown an ad in which

Bugs Bunny was depicted. Then they were asked, *"Did you shake hands with Bugs Bunny when you attended Disneyland?"*

One third said yes, they had shaken hands with that wascally wabbit. Problem is, Bugs is a Warner Bros. character—not a Disney creation at all. So Bugs had never appeared in the park. The researchers had created a false memory in their subjects merely by suggesting the idea of this impossible meeting. In another study, researchers convinced half of the participants that they had taken a hot air balloon ride that never occurred.

The functioning of an adult's memory is suspect. But childhood memories are the most highly suggestible. The most famous example is given by psychologist Jean Piaget, whose 1951 book, *Play, Dreams and Imitation in Childhood*, includes the following passage: "I can still see, most clearly, the following scene. . . . I was sitting in my pram, which my nurse was pushing in the Champs Elysees, when a man tried to kidnap me. I was held in by the strap fastened round me while my nurse bravely tried to stand between me and the thief. She received various scratches and I can still see vaguely those on her face. Then a crowd gathered, a policeman with a cloak and a white baton came up, and the man took to his heels. I can still see the whole scene, and can even place it near the tube station."

Piaget believed in the reality of this event, which he had heard as fact, till he was fifteen years old, when it was discovered that his nanny had made the story up to get a reward. In setting out to write this chapter, then, I considered the testimony of everyone older than me more reliable. And I further considered the collective picture they presented to be more important than individual accounts. In sum, when they agreed on the details I granted those details more validity.

In that sense, the Family Ghost story sails right past Piaget's kidnapping. The accounts of the close witnesses agree on the details related so far, for instance.

Over the years, as I mulled over whether or not to ever write about the episode, I had conversations with my older siblings, parents, and some other relatives.

I spoke about it to my oldest brother, Jerry, in fact, the last time I saw him before he died. He recounted the basic particulars I describe but resisted searching his memory any further than this: lots of banging over lots of nights, more stories from our sisters, the appearance of the family priest. Then he confessed that he didn't want to talk about it at all. "I don't know why," he said, multiple times, until finally he admitted, "It freaked me out."

A cousin, and my aunt, still remember *hearing* about it. They learned some of the details as they happened—including the banging, and the bit about the blessing of the house. My other brother and my sisters retain their own memories. But for our purposes, the most detailed account came from my parents, who were charged with figuring it all out. Shortly after I graduated from college, many years before my mother died, I sat down with them and talked about the whole thing. And this is the story they told me.

The banging occurred only after midnight and seemed to respond to their actions: that is, when they came upstairs, it stopped. But if they came back down too soon, it started again. "Too soon" was never an amount of time they quantified. But they did develop some standards and practices. Any time the banging started, which they described as "booming" and "incredibly loud," they took to spending an hour or more sleeping alongside us before going back to their own bedroom on the first floor.

It was not, as it were, an inside job. The kids were all accounted for, and under observation by each other during at least some of these episodes. The sound was loud and seemed to originate from the second-floor walls or ceiling, and we never succeeded in recreating it. It seems clear to me, anyway: this particular ghost story will not yield to a *Scooby-Doo* solution, the climax coming when that smart Velma tears the sheet off one of us.

My sisters shared stories with my parents of those more far-out paranormal happenings: the covers yanked from their beds, the woman who swept through their room. My parents agreed it was easier *not* to believe them. For one thing, the sound was the only strange experience they had encountered themselves. And, more pressingly, who wants to think their daughters are plagued by some paranormal force?

They dismissed my sisters' stories, then, and focused on the sound. Skeptics don't like to acknowledge this sort of thing, but not all believers *want* to think their house is haunted. My parents spent months, and more than a few sleep-deprived nights, looking for prosaic explanations: Each of us swore the banging sounded powerful, like someone trying to pound in through the roof or wall with a hammer. And yes, my parents briefly wondered if the source of the noise might be raccoons jumping on the roof. But more seriously, they wondered if it was some sound in the pipes. Many supposed hauntings can be explained by simple plumbing problems or the expansion of pipes inside the house's walls. Water hammers, or fluid hammers, typically occur when dishwashers, washing machines, or toilets suddenly stop the flow of water. When the water is shut off, there is a loud banging sound. All that energy in the flowing water needs some place to go and converts into acoustic energy. Though admittedly my parents' analysis was that of a middle-class couple, busy with the rigors of child rearing, they ultimately concluded the water or heating pipes couldn't be responsible: The sound happened at regular times, after midnight, as everyone slept, and this big, banging sound went on longer and louder, every time it occurred, than any plumping problem with which they were familiar, including the water hammer.

Eventually, feeling they had run out of potential prosaic explanations, and tired of the phenomenon, which often interrupted a precious night's sleep, they settled on a paranormal possibility. They contacted the priest of the parish we attended. An appointment was set. Big Father Crowley entered, flicking holy water and saying prayers. And that night, things got even more surreal.

I talked to my parents about this whole episode maybe twenty years after it happened. I was staying with them in Florida briefly, after graduating college. They had retired there and I would be moving to the Northeast. We had eaten lunch, cleaned up the dishes, and sat back down at the table to talk. What struck me most was *this* part of the story, not only because the details were so outlandish, but because of the way they reacted to telling it. They had shared the same general account with me in the past, as had my brothers and sisters. But this was the first time I'd led them through the telling, in as much detail as they could remember. Looking back, it was my first post-grad reporting experience. I wanted to get past the history that had accrued in my own mind, having lived around the story for so long. And that required going straight to the sources closest to it. What I learned is that sometimes the *way* a story is told is as telling as the story itself.

IIIIIIIIIIII

WE HAD STARTED OFF our "interview" almost playfully. But maybe an hour in, when I asked them to walk me through the last night, the color drained from their faces. Their bodies seemed to grow heavier as they sagged forward in their chairs. Their voices lost volume and vigor. They seemed not just to be recalling the details but *reliving* them. They looked as gut-sick and scared as they claimed to feel that night. "We didn't know what was happening," my father said. "And we didn't think we could stop it."

So, what did happen?

Well, usually, when the banging started, we waited for them to come and comfort us. But that night the thumping was so loud, so encompassing, so threatening, that everyone congregated downstairs. The lamp over the dining room table swung, side to side, with the force of the blows on the house. (I have one additional memory of my mother hollering, to my father, "Jerry, the whole house is going to come apart!")

My parents talked, briefly, about packing an overnight bag. They were going to take us to a hotel. And then, suddenly, something new happened. The noises shifted. For the first time ever, the banging lost its amorphous quality, and assumed a precise location. It hit upon the stairs.

Retelling the story, my parents fell quiet.

"What did you do?" I asked.

"We were so scared," my mother said.

So what they did was, they reached out and held hands. In fact, when they shared this detail, they reached out and held hands again, all these years later, across the dining room table. "I was in the lead, a little in front of your mother," my father said, "because I figured I had to protect everybody."

From their position, my parents could see down the hallway to the front door. The stairway banister ran parallel to that same hallway. And from their vantage point, my parents could not see the stairs themselves—or what was on them. Booms sounded from the stairway as whatever it was came down, one step—*boom!*—at a time. To my father, the sound suggested a child throwing a tantrum. But this was one hulking child. Both my parents felt tremors in the floor every time another bang sounded. And when it reached the very bottom, there was a small change in the rhythm, an extension of the patterned silence between explosions, as if the Family Ghost, the water hammer(?), whatever it was, having reached the last step, was regathering itself to hit upon the landing—and my parents' hearts—with one last, ferocious, floor-shaking effort.

And then it did.

The last sound was the loudest—a big, two-footed dismount. The noise stopped then.

And it never happened again.

||||||||||||

I RECOGNIZE HOW TRULY fantastic every word of the Family Ghost story is, and here I am, a reporter. I can hear the skeptics chortling, and I laugh right along with them. I realize that the ending of this story is particularly unbelievable. And I know for some people it will undermine my credibility. But what if I told you that I don't believe it myself? And what if I told you, further, that I don't *dis*believe it either?

The truth is, as a teenager and into my early twenties, I used to love telling this story. But over time, as the people I told it to were themselves older and more committed to a worldview, I pretty much stopped telling it altogether. Because believers left me utterly cold—through their unblinking acceptance of the story's every particular, and the way they prattled on, afterward, about the nature of poltergeists, negative spirits, and the afterlife. Here I was, thinking all this stuff was mysterious and unproven. And there they were, encyclopedias of the unknown.

Skeptics, on the other hand, greeted this story with scorn. And I was surprised to find that they delivered their prosaic explanations with greater emotion than any believers offered their more amorphous visions. The usual explanations have revolved around plumbing and the house settling. In response, I nod. Neither explanation strikes me as sufficient. Water hammers, expanding pipes, and settling floorboards don't strike repeatedly, incessantly, at that volume, over tens of minutes—or come down the stairs and then disappear forever. Further, my father says the water pipes extended only partly up into the walls of the second floor and could not account for a noise that seemed to originate higher than that, in the walls and ceiling.

One young skeptic once told me it *had* to be the plumbing. I explained why that particular prosaic explanation seemed unlikely, then said, in what I hoped was a conciliatory fashion, "But, hey, maybe it was."

"Wh-wh-what do you mean?" he sputtered. "*May*be? It *has* to be."

Taken aback by all this emotion, and from a rationalist, no less, I tried my best to ease the tension. But he seemed incredulous that I maintained the matter was simply . . . *unexplained*.

"I'm not saying it was a ghost or a spirit being," I said. "I'm not saying it wasn't. I'm saying we don't know what it was."

What I learned, over the years, was that for many skeptics, even an unlikely materialistic explanation that doesn't fit the stated facts is immensely, emotionally preferable to simply saying, *I don't know*. And so I have most enjoyed sharing the story with people who occupy a spot somewhere in the middle, who offered prosaic explanations and listened to my responses in a spirit of inquiry and conversation—in the playful spirit, in fact, that I have endeavored to write this book.

I've lived with this story for so long, in all its ambiguity, that the ongoing debate over the paranormal has seemingly always been one of my abiding interests. And yet, I also feel as if I had little choice in the matter. I was born into this, in other words—the Family Ghost is something I never, ever asked for. And so, for most of my years, I've watched all this from what might most accurately be termed the fringe of the fringe—not participating directly, but monitoring what skeptics and believers say about the possibilities. And I've looked for a prosaic explanation that does justice to my parents' version of events.

There are some intriguing ideas out there, like infrasound—sound waves too low to be audible to the human ear but powerful enough to influence our perception. Infrasound is produced by low- and high-pressure weather systems, storms, and human-made objects like large subwoofer speakers, diesel engines, and some wind turbines. The most well-known researcher in the area of infrasound and hauntings is engineer Vic Tandy, who saw a gray blob out of the corner of his eye one night and ultimately traced the experience to infrasound. He subsequently hit the road, looking for infrasound in reportedly haunted sites, with mixed success. And there are now others out there, pursuing his work and that of Dr. Michael Persinger, who believes electromagnetic waves can explain both religious experiences and ghosts.

In 2003 researchers in the United Kingdom reported an experiment in which low-frequency sound waves were played during a concert. Reports of unusual experiences increased by 22 percent during the periods in which the sound waves played, including anxiety, sorrow, chills, tingling in the spine, and pressure on the chest. But no one had visual or auditory hallucinations.

In another experiment, dubbed the "Haunt Project," organizers locked seventy-nine people, individually, inside a specially constructed chamber for fifty minutes—bombarding them with infrasound and Persinger-styled EMF (electromagnetic field) waves. Such waves are found mostly near volcanoes and fault lines. But this study found no effect from the waves or the infrasound. In fact, people reported just as many odd experiences when the EMF waves and infrasound were turned off.

Of course, there can be no one-size-fits-all explanation for an experience that has occurred in humankind for millennia, long before subwoofer speakers and the diesel engine made the scene. And so one other possibility to toss on the pile is that of a psychological explanation, beginning with "the fantasy-prone personality" or FPP, for short. Proposed in 1981 by the psychologists Sheryl Wilson and Theodore Barber, the FPP is a kind of debunker's dream. In this formulation, the fantasy prone are people who not only lead rich fantasy lives—they can actually blur the lines between fantasy and reality. Wilson and Barber identify fourteen characteristics of fantasy proneness:

1. Being an excellent hypnotic subject
2. Having imaginary playmates as a child
3. Fantasizing frequently as a child
4. Adopting a fantasy identity
5. Experiencing imagined sensations as real

6. Having vivid sensory perceptions
7. Reliving past experiences
8. Claiming psychic powers
9. Having out-of-body or floating experiences
10. Receiving poems, messages, and such, from spirits, higher intelligences, and the like
11. Being involved in "healing"
12. Encountering apparitions
13. Experiencing hypnagogic hallucinations (waking dreams)
14. Seeing classical hypnagogic imagery (such as spirits or monsters from outer space)

In other words, the fantasy prone are more likely to believe, among other things, that they experienced something that mainstream science rejects. And yes, the reasoning sounds a bit circular. Like any diagnosis in psychology, fantasy proneness lies, to some extent, in the eye of the beholder. Most people probably have some of these traits. But answering yes to any six of the fourteen questions is believed to mark one as fantasy prone. (For the record, my tally is four; I answered yes to 1, 2, 3, and 6.) But I also wonder, Could having a single strange, unexplained experience *make* someone fantasy prone? We do reevaluate our life experiences as we go along. Some events are forgotten entirely. Others seem to grow more vivid, depending on what's happening in our lives right now. Could Elisabeth Kübler-Ross, the psychologist with whom our tale first started, have been *made* fantasy prone by hearing the incredible tales told to her by her patients?

Those who knew Kübler-Ross claim they initially took her to be like most scientists—the kind of woman who considered death to be the end. It was only years after she started encountering *other information*, in the near-death experiences and deathbed encounters of her patients, that she started visiting psychics and the like. Based on what we know of her biography, if Kübler-Ross took the FPP test before her conversion, she might have scored under six. But if she took the test after, say, 1975, she would have answered yes to about nine of the FPP traits—at the least.

The FPP is perhaps not so terrifically awesome as the skeptical mind might have us believe. Skeptics love to put reports of alien abduction down to fantasy proneness. And I don't particularly blame them. I didn't cover alien abductions in this book because I thought it would be too difficult for me to overcome my own bias that abductees are fantasizing. But psychological assessments of abductees are mixed—with numerous studies finding them no more likely to engage in fantasy than anyone else.

There are other psych-based tests for irrational thinking. The Magical Ideation Scale seems particularly popular—and indeed, predictive, if also a bit extreme. The form is comprised of thirty true/false statements:

10. *The government refuses to tell the truth about flying saucers.*

13. *Numbers like 13 and 7 have no special powers.*

14. *I have noticed sounds on my records that are not there at other times.*

15. *The hand motions that strangers make seem to influence me at times.*

And, my favorite:

21. *I have sometimes had the passing thought that strangers are in love with me.*

People who choose the fantastic answer to a multitude of these questions or who report particularly detailed and dramatic personal experiences related to a subset of these questions are, statistically, more likely to develop some sort of schizophrenia-spectrum disorder. And for that matter, people who qualify as fantasy prone are more likely to have experienced some kind of trauma as a child, encouraging them to develop a rich fantasy life. And so the skeptical community has used magical thinking and the FPP as whipping sticks to suggest there is something very different and at times quite literally *wrong* with the people who disagree with them. But to me, while this whole area of research is surely fruitful, it also seems underdeveloped. In each case, both the fantasy prone and the magical thinkers often sail through life with no mental problems and exercise an increased faculty for creativity. So,

I'd love to see some studies on what might be called *sane* belief. I also find it curious that, while there is a body of research on the paranormally inclined, there is comparatively no research at all on why unbelievers hold the opinions *they* do. Is there a particular personality type associated with *rejecting* all paranormal phenomena?

Our pal skeptic, Chris French, admits the lack of research on skeptical psychology reflects simple bias. "When you look at the people who are doing this kind of research," he says, "we wouldn't think we *need* to be studied. But I think it's an important topic, and I intend to carry out some research of my own."

French cautions that the theories he holds are pure speculation at the moment. But he thinks there is probably a scale for unbelievers. "Someone like James Randi or Richard Dawkins, both of whom I hold in great respect, were probably *born* not believing," he says. "And they would represent one end of the spectrum."

As we come back down the scale, he says, we might find people like himself. But even he was once "out there" (my words) with Randi and Dawkins.

After converting to skepticism, French, by his own account, wrapped himself in dogma. He was a hardline skeptic. And in his mind, practicing psychics were frauds. Believers were idiots. "I didn't have any tolerance for ambiguity," he says. "I felt that acknowledging areas that might require more research just sort of muddied the waters."

He mellowed as the years and experience *re*opened his mind. He remains a committed skeptic. But "I do question how much we *know* in the empirical sense you're talking about. I think the best tack to take is one that's nondogmatic, in which we go on learning and remain willing to revise our beliefs."

So where does all this leave me and the Family Ghost?

Well, more than thirty years have passed, and I don't know for sure what caused my family's experiences. Yet here's the kicker: neither do you. The experience happened too long ago, and the events described don't comport with any of the prosaic explanations thus far offered. This means it's nothing for a believer to build his life or worldview around. But I also don't see it as a viable exhibit in a treatise on why ghosts don't exist.

I'd be remiss, in fact, if I didn't note that not *all* the research goes against ghosts. In 2010, Dr. Barrie Colvin, a skeptical psychologist, conducted an analysis of recorded poltergeist sounds and claimed they had a radically different acoustic signature than normal rapping or knocking. He studied ten poltergeist cases and found the same odd sound waves every time. "In each of the recordings, when subjected to acoustic analysis, a particular sound pattern is detected which so far remains unexplained," reads a Society for Psychical Research announcement of Colvin's findings. "Attempts to replicate this pattern in ordinary ways have so far been unsuccessful."

His speculative operating theory, for now, is that in poltergeist cases the sound seems to emanate not from the surface of a material—as if someone is banging *on* the wall—but from *inside* the material itself.

I also spent several long months investigating haunting claims with the now-deceased Lou Gentile, a Philadelphia ghost hunter. And I wrote about my travels with Gentile for a story published in the alternative newspaper, *Philadelphia Weekly*. My time with Gentile undermined the positions of both skeptics and believers alike. We encountered odd sounds that we could not trace to any known source (and in three instances, the noise seemed to be responding to what Gentile said). So prosaic explanations aren't always available—that is, unless we allow our commitment to the rational to make us downright irrational. But to leap from an unexplained sound to the existence of disembodied life forms is, well, too great a leap to make, based on the evidence at hand. In short, then, my travels with Gentile forced me to say *I don't know* quite a bit. And I miss him. But my hunt for an answer didn't end there.

In fact, I did find and speak with the families who lived in my childhood house after we moved.

But that is a story best left to the end of this entire tale.

6

To Infinity and Beyond

What Science and Spirituality Look Like from Outer Space

I am convinced that of all the people on the two sides of the great curtain the space pilots are the least likely to hate each other. . . . I believe that the tremendous and otherwise not quite explicable public interest in space flight arises from the subconscious realization that it helps to preserve peace. May it continue to do so!

—Konrad Lorenz, *On Aggression*

Present systems for getting from Earth's surface to low-Earth orbit are so fantastically expensive that merely launching the 1,000 tons or so of spacecraft and equipment a Mars mission would require could be accomplished only by cutting health-care benefits, education spending or other important programs—or by raising taxes. Absent some remarkable discovery, astronauts, geologists and biologists once on Mars could do little more than analyze rocks and feel awestruck beholding the sky of another world.

—Gregg Easterbrook, "Why We Shouldn't Go To Mars"

For Edgar Mitchell, a journey to the moon and back was a family trip. And the path home made the biggest impression on him. The *Apollo* spacecraft was in barbecue mode, a slow rotation, like a backyard rotisserie, designed to make sure the sun didn't overheat one side of the craft. And for the first time in many days, Mitchell had some time to sit and think, to enjoy some sense of accomplishment and look out the window. The long struggle of learning and training necessary to reach the moon was behind him. Now, the lunar surface was behind him, too.

Astronauts, like professional athletes, enjoy their greatest, most public accomplishments early in life. Mitchell, the sixth man to walk on the moon, was forty years old. At the time, he didn't know what was next. But just outside his spacecraft window laid the work that would sustain him throughout the rest of his life.

As the craft rotated, Mitchell's view shifted. He saw the Earth, the moon, the sun, and a vast field of stars, in a panorama that repeated itself every hour, with the slow roll of his spacecraft. From here, as he looked down on Earth, he could recognize seas and continents. He could identify countries and cities. He knew that his younger brother, in the Air Force, worked somewhere on the small peninsula of Southeast Asia, flying missions in Vietnam. Mitchell had flown missions of his own, in the Korean War. Thinking of this, and staring out his capsule window, he reflected on how violent life on Earth is, despite the planet's peaceful blue and white appearance. He looked again, shifting his point of focus from the Earth to the inestimable star field of which the Earth is just a part. And then, something happened.

Mitchell would spend the rest of his life trying to understand the implications. But suddenly, everything he thought he knew seemed grossly wrong.

He had, many years earlier, set aside the religious views with which he had been raised, considering them the leftover attempts of prescientific people to understand their existence. But the vision of reality that had come to occupy his mind since then, a purely scientific view, was also now shattered by what he saw. The idea that every star, every planet, every object in the cosmos, was separate and distinct had worked just fine for him. The Newtonian model of a predictable, physical universe had, in fact, provided the framework for a science that shot him into space and back. But looking out at the cosmos from space, he did not so much understand a new vision of reality as *feel* it.

He felt the Earth, the moon, the sun, and the stars. He felt his own relationship to all these things. He even felt the blackness in between them. The borders of flesh and bone disappeared. He felt the sensational tremors of his own being extending out into space. Forget distinctions between continents and countries. Edgar Mitchell now felt there were no boundaries between our bodies and the celestial bodies. He felt no distinction between himself and the nothing of black space. He suddenly experienced life—with no distinctions at all.

Our existence, he suddenly believed, is the product of an intelligent evolution—more grand than religion or science has described. This is the source of us. And we remain connected to that source. Mitchell *felt* this in what he describes as "an ecstasy of unity."

In his talks, and in his books, and in the articles written about him, Mitchell usually just speaks of this first epiphany. But when I met with him, at his home in Palm Beach, Florida, he explained that he moved in and out of this state for three days. "I went into that state maybe two or three times an hour," Mitchell said. "When it continued to happen all the way back, when every time I would look out the window I would be having a repeat of all this, I did wonder, 'What is happening to me? What the hell is going on here?' "

When he landed, Mitchell undertook research in philosophy, religion, and science to try and understand what he had experienced. By the time I met him, he was forty years down that road. And in person, he proved as serenely confident as that history might suggest. He lives in a well-kept, ranch-style home in a fairly secluded section of West Palm Beach. His closest neighbors are palm trees and open fields. He was nearing eighty when I met him. And except for a stiff gait, he seemed to be in tremendous shape for someone so close to octogenarian status. He was tall and lean. His grip hurt when he shook my hand.

He led me into an office, taking great delight in a pair of dogs that gathered at his feet when he sat down. He motioned me to sit across from him, a large tray of sliced-up vegetables and hot tea arranged on a table between us. I was a bit relieved at the warm reception. He had seemed rather wary of me when I first started making arrangements to visit. But as he explained over the course of a couple of days, experience has taught him to be careful.

He has been ambushed by moon-landing deniers—the small cadre of people who believe the entire Apollo space program was a massive hoax perpetrated on the American taxpayer and the people of the world. There is even a video of a feisty, seventy-something Mitchell ordering a denier out of his house. In the video, as the man bends over to pick up some papers, Mitchell forgets for a moment all about any ecstasies of unity—and knees him in the ass.

Mitchell is also the target of criticism from the skeptical community, who consider him an astronaut who never really came back to Earth. It is easy to see why. His post-*Apollo* studies led him into Eastern religious practices, mental telepathy, psychokinesis, meditation, and New Age healing. In the past decade, he also publicly endorsed the idea that extraterrestrials are visiting Earth. (He saw nothing of alien life in his work for NASA, he claimed, but military officials had since assured him of E.T.s' existence.)

Mitchell pronounces himself at peace with the skeptical community. He is playing, he says, for far bigger stakes than the whim of the moment. But from an outsider's perspective, Mitchell is perhaps not well known enough for his longest-running life's work—his explorations into the extremes of human consciousness.

During the course of my research, in fact, novelist Dan Brown released *The Lost Symbol*. Like *The Da Vinci Code* before it, the book is a potboiler in which the great mysteries of religion are connected by means of a vast conspiracy to the workings of government and the potential downfall of humankind. Brown incorporated research findings from the Institute of Noetic Sciences (IONS) in his plot.

IONS is, quite literally, Mitchell's brainchild. Mitchell founded IONS in 1973, just two years after he returned from space, as a means of further investigating the insights he received aboard *Apollo 14*. He chose the word *noetic*, which comes from the Greek word for "intuitive knowing"—because it captured the full-bodied epiphany he had in space. Knowledge occurred to him in an instant, he felt, and since then he has endeavored to use the means of science and logic to verify what he learned. With this as the organization's original marching order, IONS has investigated the more controversial corners of consciousness, looking for evidence of the unity Mitchell experienced.

In his novel, however, Brown writes that they found so much more: "[IONS] work had begun using modern science to answer ancient philosophical questions: Does anyone hear our prayers? Is there life after death?

Do humans have souls? Incredibly, {IONS} answered all these questions. . . Scientifically. Conclusively."

The truth is, of course, that IONS's mandate isn't supernatural; and no one has answered these questions—scientifically, conclusively. Directors at IONS are the first to admit it. Still, the publicity IONS received in the wake of Brown's book was welcome and overwhelming—including lengthy treatments on the Discovery channel, *Dateline*, and NPR. Hits at the IONS web site jumped 1,200 percent. Membership rolls and donations spiked. Book sales for IONS's current, leading lights took off, too. But Mitchell himself received virtually no publicity. He remains involved at IONS. He is listed as chairman emeritus and founder. And staff members there tell me his experience in that *Apollo* capsule remains foundational for them. But the institute that searches for unity has established a separate existence from the man who brought it into being.

Mitchell himself seemed fine with that state of affairs when I met him. But he expressed no thought of simply disappearing himself into the cloak of old age, relative anonymity, and death. "A psychic I trust," he told me, "predicted I will live to be 109 years old."

He fixed me, then, with a wry, gentle smile—the prediction seeming, perhaps, a little too good to be true, even to Edgar Mitchell. Maybe another three decades of life even strikes him as an embarrassment of treasure for one man. After all, his story already encompasses a vast swath of American history, reaching all the way from the dust of the Earth to the dust of the moon—and ultimately into the heart of what it means to be human.

IIIIIIIIIII

MITCHELL'S GREAT GRANDPARENTS, ON his father's side, wanted to start a new life after the Civil War. They traveled west in the 1870s, by covered wagon, with a few head of cattle. Railroads were not yet complete across the South and West. The automobile and the electric light were yet to be invented. The airplane was a fantasy. Yet in less than one hundred years, their great-grandchild would stand on the moon.

"I often tell people about that history," Mitchell explained, "so they can set the accomplishment of landing on the moon in perspective."

Mitchell's journey to the moon and back was the culmination, in his mind, of his family's trip west. But it also spoke to something grander than that. "I see it as an evolutionary step for man," says Mitchell, "from the water, to the trees, to land and out into space."

He even finds a mirror for his own epiphany of an interconnectedness among all things in his father's experience of running the family farm. Many nights as a child, after his family moved to Roswell, New Mexico, Mitchell heard his father rouse himself from bed in the middle of the night. As he grew up, he began to accompany him.

"One of the cows is sick," the old man would say. Or, "She's having trouble birthing. I have to help her."

Mitchell's dad was close to the animals in his care. They had names. And so Mitchell and his dad would climb into a pickup truck, turn on the headlights, engine idling low, and start rolling slowly over their property. Mitchell says his father never seemed to search for the animal in need of aid. Instead, he drove straight to it, across several acres, his headlights carving out a small sliver of light in the deep country dark. And sure enough, the animal needed his help. Mitchell never questioned how this was possible. Neither did his father. But in the wake of his experience aboard the *Apollo,* he decided his father was so connected to the animals in his charge that he knew precisely which bush to look under to find the heifer having trouble giving birth.

"Nature, I think, was my father's religion—his way of getting in touch with all that," Mitchell said. "My mother and grandmother, on the other hand, were very religious in a more literal sense."

Mitchell attended church with his mother, for a time. These were Baptist ceremonies, and the call to confession, so dramatic, made an impression on him. So did his mother, who wanted him to be an artist or a musician. But Roswell, New Mexico, was not so remote a place then as it seems today. In scientific terms, in fact, Roswell was an epicenter. And it was toward science that Mitchell felt himself pulled.

Each day, as Mitchell walked the white gravel road to school, he passed the home of America's first rocket scientist, Robert Goddard. Many years later, Roswell would be the source of many rumors and tales surrounding the purported existence of aliens. But in these days, other rumors emanated from Goddard's country home: That he moved to Roswell because he was *asked* to leave Massachusetts; that he required the isolation of Roswell to continue his top secret work; that strange machinery filled his home; and that Goddard conducted dangerous experiments, contraptions he brought out sometimes at night—and used to ignite the heavens.

Mitchell saw no evidence of any of this. But the mere presence of so eminent a scientist fired his imagination. And farm life gave him a firm grounding in the principles of engineering. The timetables to be met and the little available money didn't allow for repairmen to be called every time a piece of machinery broke down. So Mitchell learned, like his father before him, how every machine worked. He enjoyed it, so much, that at thirteen years old he sought and acquired a part-time job washing airplanes—the better to be near bigger, grander machines.

The mechanics and pilots there took a shine to Mitchell. They taught him how the planes were put together. And as he proved his intellect and maturity, they even taught him how to fly. At fourteen years old, he climbed into an airplane cockpit and flew all by himself. "I knew what it meant to be truly free," says Mitchell. "Released from the bonds of the Earth."

Edgar Mitchell wasn't going to be an artist; instead he embarked on a lifelong course of scientific education. He attended Carnegie Mellon University, in Pittsburgh, cleaning slag from the steel furnaces when money ran short. He graduated in 1952 with a bachelor of science degree in industrial management and enlisted in the Navy, knowing that volunteering would better enable him to choose his own path. He wanted to fly, and after the necessary training found himself piloting a jet in the Pacific theater.

By this time in his life, Mitchell had become a true devotee of science. Organized religion was, to his mind, merely an artifact. Religious texts were documents left behind by people who lacked the tools or knowledge to grasp life as it truly is. Mitchell earned a master of science degree in aeronautical engineering from the U.S. Naval Postgraduate School. He earned a doctor of science degree in aeronautics and astronautics from the Massachusetts

Institute of Technology. But instead of using all that knowledge in the interests of academic research, he took work as a test pilot.

Occasionally, the death of a colleague announced itself with a puff of black smoke on the horizon. But Mitchell learned to steel himself, mentally, against the risks. "I had to accept that whatever will be, will be," Mitchell told me. "I could only focus on the things I could control."

He already knew he wanted to be a part of NASA. And he considered his thirteen years of professional flying to be his apprenticeship, his training, for that prestigious institution. He made sure they were aware of his interest in joining them. And in 1966, the phone call finally came, opening Mitchell's path from the Earth to outer space.

||||||||||||

IF EDGAR MITCHELL WAS the only astronaut so moved by the view of Earth from space, we could dismiss him. But Mitchell's experience is typical.

The astronauts talked about it among themselves, initially. "You say to yourself '[Down there] is humanity, love, feeling and thought'" said astronaut Eugene Cernan. "You wonder, if you could get everyone in the world up there, wouldn't they have a different feeling?"

From space, as Mitchell experienced, only natural borders stand out. The lines we see on the map depicting mountains and oceans are all rendered, in brilliant color. But the thick black lines that divide nations are nowhere to be seen. "I think the view from 100,000 miles could be invaluable in getting people together to work out joint solutions," writes astronaut Michael Collins, "by causing them to realize the planet we share unites us in a way far more basic and far more important than differences in skin color or religion or economic system."

Astronaut Joseph Allen, a doctor in physics, mirrored another aspect of Mitchell's experience. "For several hundred years we have had a certain image of the Earth," he said. "Now an intellectual understanding is being replaced by an intuitive, emotional understanding."

The space program provided photos of our planet as a lonely round sphere, sitting improbably in a dense black field. Those images are now widely credited with fomenting the modern environmental movement and garnering funding for the Environmental Protection Agency (EPA). "With all the arguments, pro and con, for going to the moon," said Allen, "no one suggested we should do it to look at the Earth. But that may in fact be the most important reason."

The sense of unity Mitchell describes clearly manifests in astronauts as a desire to protect the Earth—and see beyond national borders. Former Republican Senator Jake Garn was aboard a 1985 space flight in his capacity as chairman of the committee overseeing NASA's funding. He felt all political borders melt away. And flying over Third World countries, he wondered why the governments of the Earth had not mobilized to feed every hungry child on the planet. "Why does this have to be?" he said, in 1986, when the Cold War was still running hot. "[Looking at the Earth from space], you realize that the Russians, the Nicaraguans, the Canadians, the Filipinos—it doesn't matter where they're from—all they want to do is raise their kids and educate them, just as we do."

For a time the stories astronauts told and the pictures they sent back to Earth seemed a gathering force for social and political change. Astronaut Rusty Schweickart founded the Association of Space Explorers, a group comprised of the handful of individuals from thirty-four countries privileged enough to have seen the Earth from space. He hoped their unity, across borders, would send a message about the kind of evolutionary shift he thought possible. In Schweickart's view, spaceflight might lead to a planet bound by its common humanity, not identification with a particular flag.

Author Frank White, to whom I am deeply indebted, interviewed thirty astronauts about the life-altering view from space. He dubbed the phenomenon the "overview effect" and wrote about it in a book of the same name. He found many astronauts had switched career paths and made new, surprising decisions after their trips into space. His book details several fundamental shifts in their thinking, all redolent of Mitchell's "ecstasy of unity" and the idea of intuitive knowing.

Astronauts, he found, don't just understand intellectually that political, religious, and cultural boundaries are purely human-made constructs. Once in space, they viscerally feel it; and they place their own self-images into a far larger framework. The astronaut Schweickart, for instance, said that he understood himself as the "sensing element for man." Just as an individual's fingers reach out and trace the contours of some object, he was, he realized, taking in data that would be relayed to all humankind.

White also found the astronauts, like Mitchell, reframed their view of the Earth. In what he calls "the Copernican perspective," the Earth is seen as just a part in the whole of the solar system. And more broadly, the Earth and its solar system are felt and understood as a mere part in the whole of the universe.

Any momentum for change brought about by these insights, however, quickly stalled. The American space program has gone through a long down period, as taxpayers urged government to spend their money on more practical, Earthbound issues. But there are perhaps other, more intriguing reasons that this country lost its taste for space.

The majority of us are so content with the worldviews we've established for ourselves that we can be gifted with some profound insight, from someone else, and summarily forget, dismiss, or denigrate it. This is understandable. We are busy people, with busy lives. Reframing our understanding of the world 'cause Rusty Schweickart said so, or because we got some pretty pictures from space, is not something we're likely to do. So whatever energy for change the space program produced naturally dissipated itself in the hustle of our days. Further, scientists have made no move to study the overview effect. And this, too, is understandable. Scientists are conditioned to disregard anecdotal stories—and so we're stuck in a merciless Catch–22: until we get repeated studies of the overview effect, quantifying its impact on the astronauts who experienced it, we're unlikely to see any studies at all.

The astronauts themselves are painfully aware of the problem, comparing their collective experience to one drop of dye in the vast, vast sea of human endeavor. From an educational perspective, it seems, what the astronauts are confronted with is the gulf that lies between direct experience and what I call the "mere knowledge" that comes from other forms of learning. Reading about how to hit a baseball, in short, is no substitute for going outside and taking some swings. So you or I can claim to empathize with the astronauts and imagine how the trips into space altered their worldviews; and scientists and philosophers can claim an intellectual grasp of everything a Rusty Schweickart says. But none of us experienced it, none of us *felt* it—and felt, in turn, the experience become a part of who we are.

The astronauts themselves trained for months and years before they went into space. They knew every detail of the craft they would occupy and the instruments under their control. They were prepped, too, about what they could expect to see beyond their spacecraft. And after the space program began in earnest, they even had photographs of the Earth, hung in inky blackness, like a pendant on black velvet. But in the end, seeing our home planet from space—not in a photograph, but with their own eyes—proved to be transformative, like standing upon a territory after seeing it represented by the lines and shadings of a map. There was, they told Frank White again and again, *no* comparison between the indirect experience they had of space travel and *being there*.

As the sensing element for man, then, they were tasked with telling us about an experience that had altered their worldviews and shifted their self-images. But we, their audience, had no real frame of reference. We got the EPA and Earth Day out of the deal. But we also got the BP oil spill and climate change, just the same.

In light of the experiential chasm between them and us, some have claimed the astronauts' descriptions are somehow not evocative enough—that we should instead send poets into space. But reading over the accounts given by the explorers themselves, I often find them so moving that I think a talent like the poet Maya Angelou's could make a difference, but probably not a dent, in our own naturally thick skulls.

Mitchell's own rendering of what he likes to call his *a-ha* moment, his epiphany, is remarkably vivid: "Billions of years ago, the molecules of my body, of [my fellow astronauts' bodies], of this spacecraft, of the world I had come from and was now returning to, were manufactured in the furnace of an ancient generation of stars like those surrounding us. This suddenly meant something different. It was now poignant and personal, not just intellectual theorizing. Our presence here, outside the domain of our home planet, was not rooted in an accident of nature, nor the capricious political whim of a technological civilization. It was rather an extension of the same universal process that evolved our molecules."

Mitchell is sometimes seen as a party of one. Other astronauts found the experience affected them spiritually, but only he publicly used the view from space as a gateway to exploring the paranormal. White, in speaking to as many astronauts as he could, however, ultimately concluded that it was Mitchell's dedication to the overview effect that truly set him apart.

"When we talk about the astronauts and their experiences, most of them were changed in some way," White told me. "But I don't know of anyone who has, to the degree Edgar has, taken the experience itself, and spent their life trying to work through the implications of it. Edgar has relived it so many times, in so many ways. He has written about it, thought about it, spoken about it, again and again. He founded IONS to try and understand it. He just keeps working through the implications of it. And what impresses me is the rigor with which he has done it."

Mitchell puts it all more succinctly: "I've endeavored to follow the path the experience suggested," he says, "to see where it leads."

IIIIIIIIIII

BECAUSE IT WAS HIS own consciousness that provided Mitchell the sense of being one with all creation, and because no one has defined the mechanism that makes consciousness possible, Mitchell felt his course was clear. It was toward the riddle of consciousness that he would first look for evidence of the unity he felt. Forty years later, he is still endeavoring to mine those same depths—sometimes in a manner dangerous to his reputation.

Shortly after his return, in the early 1970s, he struck up a relationship with the famous Israeli psychic Uri Geller. Mitchell stood by as Geller took part in a series of experiments at the Stanford Research Institute, experiments in which Geller purportedly read minds, viewed remote images with his "mind's eye," and attempted to bend spoons, psychically. I write that he "attempted" to bend spoons because Mitchell is the first to point out that, under strictly controlled conditions, Geller failed to bend any spoons at all. But reading minds? Remote viewing?

"Yes," says Mitchell. "He did that."

Today, the Internet is filled with claims and counterclaims about Geller—that he is a con man, a magician, a genuine psychic, or some combination of them all. I spoke to Geller on the phone and exchanged emails with him and found him to be difficult at best. "Uri," I said, by way of introduction, "I'm one of those people who would perhaps like to believe but find myself unable—"

"I don't give a damn what you believe," he hollered at me over the phone, his first words since *Hello, Steve*. "Why should I care what you believe, or if you're too stubborn to believe in anything at all? I don't care. I'm way past trying to prove anything to anyone!"

"Whoa, whoa, slow down, Uri," I said.

By this time, we had already exchanged emails in which I had assured him that the thesis of my book did not require me to make fun of Edgar Mitchell—the exact opposite, really. I was advocating we not make fun of anyone. Reminding Geller of this, I calmed him down enough that he went into one of his time-tested rants thanking the skeptical community. "They made me very wealthy," he said. "They brought me more attention, for free, than I could have paid the greatest advertising firm on Madison Avenue to get."

In the end, I felt like Geller taught me nothing at all, sharing only platitudes about Mitchell, who, for his part, merely smiles at the more colorful aspects of Geller's act. Mitchell's exact position on Geller is a bit complicated: he agrees with the Stanford Research Institute's conclusion that Geller's metal-bending ability could not be proven scientifically; but he also believes, on a personal level, that Geller bent spoons by means of psychic power. He saw him do it too many times, he said, under conditions he found stringent enough to constitute that kind of personal proof. Perhaps, he suggested to me, the conditions under which he failed to perform were . . . too rigorous, impeding Geller's ability.

If you listen closely, of course, you can hear the skeptics laughing. But Mitchell has never embarrassed himself in the manner Elisabeth Kübler-Ross did, as recounted in chapter 1. The world-famous psychologist, Kübler-Ross, was forced to abandon the psychic she'd hired to work at her institute.

But Mitchell has never publicly aligned himself with any particular mystic for any great length of time. And even his position on Geller allows him an intellectual out: he considers the psychic's claims of metal-bending ability scientifically unproven. Still, the range of personal, strange experiences Mitchell has claimed runs long: he spent much of the 1970s meeting with a variety of healers and psychics. Most failed to impress him. But a few did.

The most compelling experience he reports relates to his mother, and it's easy to see why he finds it telling: the little drama that played out between them, if true, illuminates a nexus point among the paranormal, belief, disbelief, and consciousness.

Mitchell's mother had been losing her eyesight for years, due to glaucoma, and was legally blind without her glasses. Just beginning his explorations into the claims of healers, Mitchell decided to introduce his mother to a man named Norbu Chen. A Buddhist and self-proclaimed Tibetan shaman, Chen agreed to try and heal Mitchell's mother and restore her eyesight.

The three gathered in a quiet hotel room. Chen sang a strange mantra, passing his hands slowly over the head of Mitchell's mother and pausing at her eyes. The whole process lasted just a few minutes. Nothing happened immediately, but the next morning, at 6:00 A.M., Mitchell's mother came rushing to his room. "Son," she said, "I can see!"

She then proceeded to read for Mitchell, unaided, from the Bible. Then she made a show of tossing her glasses on to the hotel room floor—and breaking them with the heel of her foot.

Mitchell relates this story, and many other odd tales, in his book *The Way of the Explorer.* "I am not," he writes, "by this account nor with any other anecdotal story, attempting to convince the doubtful. That can only happen when the open-minded skeptic sets out for himself or herself to view (or better, to experience) such peculiar phenomena (at least peculiar to the western mind)."

Chen's healing of his mother, he concludes, "wasn't science, but as far as I was concerned, it indicated where I personally needed to probe more thoroughly."

Remarkably, according to Mitchell, his mother went about her daily routine for several days after this healing without her glasses. Her vision was restored. Then Mitchell's phone rang. His mother wanted to know if Chen was a Christian. She herself had remained a fundamentalist, so Mitchell wanted to keep the truth from her. But being a good son, he couldn't lie to his mom.

He could hear the disappointment in her voice. And within hours, her eyesight deteriorated. She needed thick eyeglasses again to see at all. For Mitchell, this was another anecdotal story that spoke to the power of a person's belief system to effect not only what information they would accept as valid but their own health. It was a story that spoke to the power of consciousness.

What Mitchell and his mother may have encountered is a particularly dramatic example of the placebo effect. The belief we're being healed is often enough to successfully diminish symptoms like pain or the breathing difficulties associated with asthma. It is usually stated that the placebo effect seems to have no power over the illnesses that underlie our symptoms. But researchers have demonstrated an increased interest in seeing just how far the placebo effect, or belief, can take healing. There are occasional reports of impressively dramatic effects, including the well-documented, curious case of a cancer victim.

In the *Journal of Projective Techniques,* it was reported, a Mr. Wright was in the end stages of cancer. His body was riddled with tumors. His lungs were filled with fluid. He needed an oxygen mask to breathe. His doctor was about to go home for the weekend, expecting his patient to be dead by the time he returned on Monday. But Wright heard that his doctor was conducting research on a new cancer drug called Krebiozen. He begged to receive the treatment, and his doctor relented.

Two days later, Wright's tumors shrank by half. He was the only one who seemed to be benefiting from the introduction of Krebiozen. But his doctor didn't tell him that, and he continued injections for the following ten days. Wright went home healthy.

Two months later, however, Wright heard a report that the drug was, thus far, proving ineffective. He immediately fell ill, his tumors returning. His doctor, seizing on the placebo effect, lied to him. He told Wright they had a newer, double-strength version of the drug that he believed would get results. In reality, he injected Wright only with sterile water. But again, Wright's tumors disappeared and he returned to his normal life. Unfortunately, he later saw continued newspaper coverage describing the dramatically unsuccessful tests on Krebiozen. He immediately got sick again, was admitted to the hospital, and died.

IONS became Mitchell's scientific arm for attempting to understand such happenings. And IONS's Remission Project, in fact, documented 3,500 cases of diseases suddenly and inexplicably retreating, culled from eight hundred journals, written in twenty different languages. To medical science, such cases have always been viewed as happy curiosities. To Mitchell and the scientific staff at IONS, they suggest a whole course of research into the power of mind and consciousness. And while an entire book could be written probing the pros and cons of their work, allow me to summarize it concisely: IONS has uncovered the same sort of contentious evidence for mental telepathy that has been found in research labs all over the world—a so-called psi effect that appears so small as to be without practical application, but so interesting, scientifically, that it may speak to the unity of mind and matter Mitchell felt aboard his spaceship.

Their findings are, of course, not embraced by the larger scientific community. And until they are, nothing figures to change. The skeptics will cry, like policemen guarding a murder scene, *There is nothing to see here, please go about your business!* But I like IONS, if only for the incredible conclusion it could provide to the story of Edgar Mitchell. Because if (or when) this shift ever happens, if this small psi effect is ever accepted, it might mean that Edgar Mitchell will be remembered not so much for his walk on the moon, but for his experience on the ride back. He will be remembered, most fondly, for what he accomplished after he hung up his spacesuit.

In Seattle, when I was researching my chapter on telepathy, I asked the statistician Jessica Utts what she thinks the most promising field is for those who do believe in psi. "If there is one test that might really convince the majority of scientists," I asked her, "what would it be?"

We were gathered in a small hotel ballroom, at a conference of parapsychologists. And in response, Utts nodded across the room at Dean Radin, the leading researcher at IONS, who sipped a cocktail and laughed with colleagues as we talked. "I think it's the precognition work Dean is doing," she said.

Radin has written extensively about a number of experiments, conducted by himself and other researchers, demonstrating that a measurable physiological effect can be observed in the moments before a subject is exposed to emotional stimuli. Given the long-running debate over psi and his own involvement in it, hearing that his institute might hold the inside track on settling things might be seen as a point of pride for Mitchell. But the truth is, he seems blithely unconcerned with whether or not IONS's findings are embraced any time soon. This may in part be due to the meditative practice he has maintained ever since his capsule ride.

Mitchell's experience of unity is in fact something meditators can feel just by engaging in regular practice on their living room floors. But Mitchell's come-what-may attitude about the skeptics also reminded me of what he said about learning to live with the danger of being a test pilot. In looking into the paranormal, Mitchell seems to have adopted the same philosophical stance he did when he strapped himself into an experimental plane. "It would be nice for our work to be accepted," he said. "But the workings of science, of what knowledge is embraced and what is rejected, are sociological. That's not something I can control."

Predicted to live to the ripe old age of 109, Mitchell seems content to let the wheels of science grind slowly on. In the meantime, he keeps working. And he has developed what he calls a "dyadic" model of the universe. A dyad is a group of two—separate but one. I listened to him talk about his idea for a long time, and readers with an interest in the subject should consult my Notes and Sources. But as it relates to this book, I like his model mostly as a metaphor, for the germ of a worldview that might be worth clinging to: men and women, Republicans and Democrats, believers and unbelievers, sports fans and non-sports fans, musicians and engineers, dyads, separate but one, all trapped on the same dusty rock, acknowledging our differences but understanding we're all connected, on a planet without borders, a planet that looks awfully improbable, awfully fragile, from the point of view of science—or the more dramatic view from a spacecraft.

When he finished describing his model of reality, Mitchell stood. "There is something I want you to see," he said.

He walked over to a closet in the hallway outside his office and came back holding a large plastic bag. When he was still some distance away, I wondered if he was going to display samples of moon rock and moon dust—an astronaut's pirated booty. But as he drew near I realized that the bag was filled with bent spoons. "I got these," he said, "during my visits with children, in the '70s, mostly in San Francisco and California."

I removed some of the spoons from the bag. A couple were bent at right angles. Others were twisted into curlicues, the bowl end of the spoon wrapped around the shaft in tight spirals. In the wake of the publicity he received for his work with Uri Geller, he explained, he received phone calls from mothers around the country who claimed that their kids had begun bending spoons like the Israeli psychic. He visited some of those nearest to him, on the west coast, to judge for himself.

"Go ahead," Mitchell told me. "Try to bend one."

I thought he wanted me to bend one of the spoons just by thinking about it, and I looked at him quizzically.

"With your hand," he advised.

I held the shaft of the spoon in one hand and tried bending the bowl end, with all my might, with the other. I pushed. I pulled. I strained. But I couldn't bend it.

"How did you get these?" I asked. And over the next few minutes, I questioned Mitchell, trying to figure out how the children had duped him.

He came in, he said, to a child's home. And he brought his own spoons with him. He sat down on the couch and asked the child to sit next to him. Then, holding a spoon upright in his hand and never letting go, he allowed the child to rub the spoon with one finger. After a few minutes the spoon seemed to go soft and pliable, like rubber. And the child simply bent it, quickly, and with no more effort than it takes to fold a straw. The spoon never left Mitchell's hand. But when the child stopped and Mitchell tried bending it himself, it was again hardened steel. He couldn't move it. He carried out this same, nonscientific experiment a dozen times or more, finding children who seemed to mimic Geller's abilities.

After Mitchell told me this story and put the spoons away, we parted for the day. But before I left, I felt compelled to make a kind of confession.

"I have to tell you," I said. "I respect you, but . . . you do realize: I just can't believe those spoons were bent by people using their minds."

Mitchell smiled at me gently. "Yes," he said. "I realize that."

Then he walked me to the large lawn in front of his house and watched me get in my car, his hands in his pockets, his face still smiling, his dogs dancing in circles at his feet.

As I pulled away, I watched him diminish in the rearview mirror, until he was just a dot of humanity on a spheroid planet, in an elliptical solar system, in a sweet spot known scientifically as the galactic habitable zone of the Milky Way.

IIIIIIIIIII

MITCHELL'S STORY IS NOT yet over, not only because he is still lecturing and writing, but because we as a society have not yet processed the meaning or implications of space travel.

We may yet get our chance.

The experience of all the astronauts who have ventured into space, and White's book on the subject, have inspired the formation of the Overview Institute. The institute is headed up by David Beaver, a magician, who might best be described as a lifelong student and Renaissance man.

His bio includes studies in nuclear engineering, physics, the sociology of perception, and the philosophy of science. When I spoke to him, he said that in addition to promoting the overview effect, he is also putting together a virtual reality stage show, completing a book on the cognitive science of magic, and consulting with the growing space tourism industry.

The combination of the Overview Institute and space tourism could in fact be exactly what it takes to reframe our understanding of Edgar Mitchell—and space travel. Virgin Galactic is booking civilian flights into space right now. Flights aboard a specially designed, carbon-composite spaceship—which boasts greater strength and far lighter weight than a standard airplane—are scheduled to begin, at the time of this writing, in 2012.

The company's sales pitch never mentions the overview effect by name. But most of its narrative is built around the view: the look of Earth from space is something seen in "countless images," according to Virgin Galactic's promotional material, "but the reality is so much more beautiful and provokes emotions that are strong but hard to define. The blue map, curving into the black distance is familiar but has none of the usual marked boundaries. The incredibly narrow ribbon of atmosphere looks worryingly fragile. What you are looking at is the source of everything it means to be human, and it is home. . . . Later that evening, sitting with your astronaut wings, you know that life will never quite be the same again."

This description is both spot on and probably a bit tame. But that is to be expected. Telling people, as Edgar Mitchell has, that "*your flesh and bone will melt away as you feel yourself become one with the universe!*" is probably not the best way to sell tickets. Telling people that astronauts are profoundly affected, usually for the rest of their lives, and that some change their vocations altogether, also qualifies as a tad too intense for a marketing handout. But Beaver's contention is that the tourism industry—and the rest of us—need to be prepared.

"I think what's kept the overview effect from having a larger impact on society, is that so few people have experienced it," Beaver told me. "And those people are generally one small part of society. They are space explorers. When more people are going into space, from various walks of life, change is going to start happening, really fast."

This sounds a bit dramatic. But is that because the power of the overview effect is overstated? Or because, when we listen to this story, we don't have ears that are prepared, in the least, to really hear it? This is the question—and the answer may be that we don't fully understand what travel into space will mean for us; that in fact we can't understand it until it happens.

Frank White, who literally wrote the book on the overview effect, is also a part of the Overview Institute. And he thinks of space travel largely as Mitchell does—an evolutionary step. "If fish could think at our level of intelligence," White said, "back before humanity existed, and some fish were starting to venture up on land, a lot of them would be saying, just as we do now about space: 'Why would we want to go there? What's the point?' And they'd have literally no idea of what venturing onto land was going to mean."

The move from water to land, according to White, is a kind of mirror in history—a pane of glass for us to stare through and understand that our next shift, from Earth to space, will be equally important. But what species has ever understood its own evolutionary future?

Because we are more intelligent than the fish, because we have developed the scientific method, because we can create art and films that provoke our imaginations, we of course have a better opportunity than a trout to understand what's next for us. But there is abundant evidence that we don't even understand what is happening right now.

The space tourism industry has long been plagued by a phenomenon known as "the giggle factor." In short, when people hear an idea that is profoundly disturbing—like the destructive effects of climate change—or scientifically challenging—like the idea of microscopic life once was—we giggle. And in the case of space tourism, we giggle because the idea seems too far-out—too remote from our experience. Our natural hubris, it seems, is most clearly captured in our automatic inclination to laugh at information we don't understand.

The implications of this for the paranormal are obviously great. Yes, some people want to embrace every New Age idea. But others laugh, just as automatically, before even considering what they're laughing at. "In my conversations with people in the aerospace industry," Beaver told me, "they expected they would announce flights into space—and that would be that. People would start calling for reservations. But it wasn't like that, and they realized the 'giggle factor' was to blame. They needed to do more work, just convincing people this is real."

That work has since been done, and what Richard Branson is selling through Virgin Galactic is real. We as a species just had (and some still have) a hard time believing it. Branson's newly designed craft have been making successful test flights, and industry observers believe that even the most far-out plans—like Bob Bigelow's idea to sell $8-million weekends in a space hotel—are *when*, not *if*, propositions. Branson even has competition, from the likes of PayPal cofounder Elon Musk. Branson has been working with no less a visionary than Burt Rutan, whose ideas can be found in the ever-so-practical unmanned drones currently hunting terrorists in Iraq and Afghanistan.

Check the clips covering the progression of the aerospace industry, and it appears that circa 2005 everyone involved adopted the same mantra: "The giggle factor is gone," as a means of overcoming the public's doubts.

When it comes to perception, perhaps just saying it can really make it so. People are now booking flights with Virgin. But according to Beaver, aerospace insiders aren't taking any chances. They are now loath to say too much about the overview effect, for fear of reigniting our mirth. And they probably aren't wrong.

My favorite example of the skepticism that greets the overview effect is a 2007 article in *Wired* magazine, covering the Overview Institute's first conference. "Scratch a Space Nut," reads the unwieldy headline, "Find a Starry-Eyed Hippie."

I think, in the end, this is the conundrum of Edgar Mitchell. He helped usher in what appears to be the next step in human evolution: from land to space. But he can only be viewed through the glasses we're grinding today. Too much of what he has involved himself in provokes our laughter—from mental telepathy, to shamanistic healers, to the overview effect itself. But where he has landed is more nuanced than all that.

"There is a strict idealism," he told me, "which is where the New Agers are at, which says that essentially there is no matter or that matter is irrelevant. And then there is extreme scientific reductionism, which tends to dominate science, and says that consciousness is epiphenomenal—a kind of illusion produced by materialistic processes. I say somewhere there is reality in all this. And our task is to find it. But both of these, the New Age way of looking at things, and the strictly materialistic way, are wrong. It's a more inclusive view of reality that I'm after."

Mitchell's journey, then, goes both in and out—as deeply into ourselves as we can go, as deeply into and as far from scientific dogma as we can get, as far out into space as we've dared. And he has paid a price.

The most poignant conversation I had about Edgar Mitchell was with Dr. Marilyn Schlitz. The current president of IONS, Schlitz has known Mitchell for close to twenty years. And I was frankly a bit afraid to share my final observation with her. "Edgar struck me," I said, "as lonely."

Schlitz was quiet for a couple of seconds before she responded, long enough for me to wonder if she was offended on Mitchell's behalf. But then she spoke: "I think Edgar *is* lonely," she said, and from there she painted a portrait of Mitchell, suggesting history will judge him, as it judges all things, far more accurately than the present can.

"He grew up on a farm," she observed, "and he has great respect for his family history. But he traveled to the moon and back. His experience of life is so unique. You have to consider: Even his fellow astronauts, there is a fraternity there that means a lot to Edgar. They had the same experience as him. But they didn't pursue it the way he did: So there isn't anyone on Earth he can look at, and feel that sense of total, shared experience. There isn't anyone on Earth who really understands him."

7

THE OPEN MIND

How New Science Is Revealing the Power of Meditation and Prayer

And when we could go no further, and were drowning on a desert, we raised our flag to follow the breath of God? But it was blowing every which way.

—Joe Henry, "Flag"

As we discuss these issues, let each of us do so with a good dose of humility. Rather than pointing fingers or assigning blame, let's use this occasion to expand our moral imaginations, to listen to each other more carefully, to sharpen our instincts for empathy and remind ourselves of all the ways that our hopes and dreams are bound together.

—Barack Obama, January 12, 2011

Dr. Andrew Newberg was late for class, as he would be every week. About ten minutes after the scheduled start time, he hustled around the corner, walking fast, a sheepish smile playing out underneath his thick mop of hair. He dressed in well-worn brown shoes, nondescript slacks, and a short-sleeved button-down shirt that hung loosely from his bony figure. "Hi, hiya, hi," he repeated to the line of students arranged pell-mell, sitting and standing on the floor around the locked door to his classroom.

These kids must have ranked, statistically, among the brightest in

America—freshmen at the University of Pennsylvania (Penn), many of them in their first semester. They smiled back at their professor. Newberg unlocked the door to the classroom. And everyone piled into the small, theater-like space. The whole room worked by means of electronics, with dials on the walls like the bridge of a movie spaceship. Newberg never did learn how to dim the lights himself. And over time, the class would seem to love him for it, smiling as they might at the eccentricities of a beloved uncle. But on day one, most of them had no real idea who Newberg was. They had enrolled in RELS 102: *Science and the Sacred*: *Neurotheology*, without knowing what they were in for at all.

Dr. Andrew Newberg works, full-time, as a radiologist at Penn, one of the nation's most prestigious teaching hospitals. But he is best known as the lead author of five books, which renders him an authority to anyone investigating the relationship between science and religion. His field, neurotheology, is, simply, the scientific investigation of the relationship between brain function and spiritual experience. But of course, in the context of our culture, there is nothing simple about that. And so Newberg's field has caused something of a ruckus among believers and unbelievers alike.

The foundation of Newberg's credibility is that there is nothing faith-based about his science. Medical imaging devices allow him to monitor the brain activity of believers as they engage in spiritual practice. What makes his work so controversial is the manner in which people choose to interpret his data: some see God in every grain of sand and every neuron; others see the brain at work and figure that's all there is. These days, in particular, it's not overly dramatic to say this is one debate that truly rages. And Newberg is the man in the middle—religious fundamentalists to one side, New Atheists to the other.

The religious, if they pay attention to such science at all, portray the findings of neurotheology as illuminating the relationship between soul and flesh. The leading New Atheists—Richard Dawkins, Dan Dennett, Sam Harris, and Christopher Hitchens—mean to put religion to as swift a death as they can manage, or, at least, to mock it into ever-narrower corners. As a result, New Atheists and materialist philosophers tend to see Newberg's findings as further proof that religious experience is reducible to mere brain function. Newberg, for his part, merely wants to find out what effect religion has on the human brain—to see how God, or the notion of God, occurs in our neurons. And his origin story, which he shares with his class, describes the sudden formation of a brand-new field of science.

Newberg's initial research subject was Robert, a Buddhist and an experienced practitioner of Tibetan meditation. The plan Newberg and his co-researcher hatched was novel in scientific terms. Newberg and Robert sat several feet apart, separated by a closed laboratory door yet connected by a single strand of twine. While Newberg sat on one side of the door and waited, Robert sat on the other, meditating. The activity of meditation has always been supremely difficult to study. The experience of meditating is purely subjective—and happening only in the practitioner's mind. A thought can't be pinned to a microscope slide, but science had advanced enough by this time, in the mid-1990s, that Newberg could take a picture of Robert's brain *as* he meditated. Still, it wasn't going to be easy. The system they had devised required precise coordination of elements as primitive as a length of twine and as advanced as a massive brain-imaging device. The experiment also depended upon Robert achieving a delicate mental state with distractions all around, including an intravenous line threaded into his arm.

Robert's goal was to reach a bliss that humans have been chasing, and finding, for thousands of years: the transcendent experience. In this condition, the human mind, normally so noisy with the worries of the day, quiets to a hush. Time and space drop away. The meditator feels one with the universe—every atom of *every* body, all part of his body. For centuries, mystics have described this root experience in varying terms, and in metaphorical language. The "ecstasy of unity," as Edgar Mitchell put it in the previous chapter, is both real and ineffable—an experience beyond words.

Newberg waited for an hour, unsure if the plan would work, and then—he felt it: a small tug on the length of twine running between him and Robert. This was the signal Robert was to give just before he reached the state Newberg wanted to study. Newberg waited a few beats, allowing Robert to achieve whatever nirvana he'd won for himself, then jumped into action. He opened the door between him and Robert and injected the intravenous line with a radioactive tracer. If the injection was precisely timed, the tracer would document the blood flow patterns in Robert's brain at the moment his meditation reached its peak.

Rousing Robert from his meditation, Newberg then hustled him to a room in the Nuclear Medicine Department. He laid him down on a long metal table and slid him under a huge, high-tech SPECT (Single Photon Emission Computed Temography) camera, designed to detect radioactive emissions.

Newberg didn't know whether this part of the experiment would work. No one had ever tried this before. But the results were all he could reasonably have wanted. Looking over the SPECT scan, Newberg could see that the areas of Robert's brain associated with judging distances, angles, and depths—in short, his position in space—had gone whisper-quiet. During normal consciousness, this area—the posterior superior parietal lobe—lights up on a SPECT scan with the furious red of active blood flow. This part of our brain has a lot of work to do. It keeps us from running into walls and missing the chairs we intend to sit in. Even when we're still, in fact, this area of the brain remains active: always aware of which parts of our body are in contact with the chair, and which are floating in space; how far away the water glass sits on the table, and how high. But in Robert, during the peak of his meditation, the blazing red turned cool green and blue. The suggestion was obvious: Robert felt himself become one with the universe because the part of his brain that tells him where his body begins and the objects around him end pretty much shut down.

Newberg studied eight Tibetan meditators and took similar pictures. Then he moved on to Franciscan nuns, who practice a form of meditation called "Christian centering prayer." A new field of science was born. And as Newberg accumulated data, he made an important finding: "The altered states of mind [our subjects] described as the absorption of the self into something larger were not the result of emotional mistakes or simple, wishful thinking," writes Newberg in *Why God Won't Go Away*, "but were associated instead with a series of observable neurological events, which, while unusual, are not outside the range of normal brain function."

In short, the world's mystics have not been kidding themselves—or crazy. But what did this say, if anything, about God or spirituality?

In most classrooms, with most teachers, the first day of the semester is an easy ride through the syllabus. But in Newberg's class, this most fundamental question of man's existence, asked for millennia—*Is there a God?*—was first-day stuff. Newberg told his class about Robert and landed them all in this contentious territory. I sat in the back of the room, watching Newberg's students turn serious with the weight of the subject.

In a strict scientific sense, Newberg has always been pretty humble about his data. His findings, he cautions, do not comprise evidence that God exists. Then again, his data also do not suggest that God or spiritual experience is simply a delusion. "We have to be careful," he said, "about how much we reduce spiritual experiences down to brain function, because it can be very complicated. If God or a sense of God is strictly produced by the brain, then all theological questions fall away. If the brain is *accessing* God, or some higher reality, then theology obviously comes back in."

The problem, Newberg told his students, is that "I can't tell you definitively what the answer is. No one can."

He used a brief analogy, versions of which regularly appear, as a kind of neurotheological disclaimer, in all his books and talks: "The brain mediates all our experience," he said. "Real and imagined. If I took an imaging scan of someone eating apple pie, certain areas of the brain would light up. Does that mean apple pie is just a delusion produced by the brain?"

In short, Newberg said, "We can't tell you the origin of the experience. But we can tell you the brain does appear to be built to have these experiences. There are examples of people reaching similar states, spontaneously. But for the most part, it takes work. Meditation and these powerful prayer experiences require dedication and practice. But people have figured out how to do this, and the question is, 'What is the source of that experience?' The answer is, 'We don't know.' Science doesn't really have an answer for you."

Newberg's students shifted uncomfortably in their seats, some with nervous smiles. I would imagine that their brains were accessing the series of complicated questions Newberg had raised: Have the world's various religions found ways for the human machine to trick itself into experiencing something that feels profound? Or have the world's religions found ways to access something real? For that matter, is there any meaningful distinction between a *truly* profound experience and one that only *feels* that way?

After allowing this heavy silence to persist for a few seconds, Newberg went on: "Here's what I can tell you," he said. "I can tell you, when people are having a particular spiritual experience, which parts of the brain light up. That's it. That's all I can tell you."

I followed Newberg's class throughout the semester, and as the weeks wore on Newberg's students told me they admired their professor for avoiding big pronouncements, for avoiding the kinds of statements that fuel opposition and debate. But the irony is that, in his unrelenting humility, Andrew Newberg might be making the biggest statement of all. And his pictures of the human brain, engaged in a spiritual quest, might be just what we need to quiet this cultural war.

IIIIIIIIIII

THE SPEECH OF RELIGIOUS fundamentalists is filled with judgment and fantasy. Just consider Pat Robertson's take on Haiti, in the direct aftermath of an earthquake that killed more than 200,000 people in 2010. "They were under the heel of the French," he said. "They got together and swore a pact to the devil. They said, we will serve you if you'll get us free from the French. True story. And so, the devil said, okay it's a deal. . . . Ever since, they have been cursed by one thing after the other."

The actions of religious fundamentalists are often as profane as their speech, and for evidence we need look no further than the Catholic Church's decades-long cover up of child sexual abuse perpetrated by some of its priests.

In response to such bad behavior, which has accumulated with the centuries, a movement has risen up—with great vengeance and furious anger.

"Violent, irrational, intolerant, allied to racism and tribalism and bigotry," writes Christopher Hitchens, in his bestselling rant, *God Is Not Great*, "invested in ignorance and hostile to free inquiry, contemptuous of women and coercive toward children: organized religion ought to have a great deal on its conscience. . . . With a necessary part of its collective mind, religion looks forward to the destruction of the world. By this I do not mean it 'looks forward' in the purely eschatological sense of anticipating the end. I mean, rather, that it openly or covertly wishes that end to occur."

Daniel Dennett writes, in *Darwin's Dangerous Idea*, "The kindly God who lovingly fashioned each and every one of us and sprinkled the sky with shining stars for our delight—*that* God is, like Santa Claus, a myth of childhood, not anything a sane, undeluded adult could literally believe in."

Perhaps most famously, the Pope of the Godless, Richard Dawkins, sounded a call to arms with an editorial he wrote after the September 11, 2001, terrorist attacks on New York and Washington, D.C. "Many of us saw religion as harmless nonsense," he observes. "Beliefs might lack all supporting evidence but, we thought, if people needed a crutch for consolation, where's the harm? September 11th changed all that. Revealed faith is not harmless nonsense, it can be lethally dangerous nonsense. Dangerous because it gives people unshakeable confidence in their own righteousness."

Newberg might seem a physically and temperamentally slight figure to hold the center of a cultural maelstrom. He is an apologetic believer who has admitted he suspects there is some higher truth in religion but further admits he cannot prove his position. He is physically unimposing. He wears a perpetual smile. He is gentle and mild. And he altogether lacks the capacity, in my experience of him, to reach the verbal extremes that excite debate. He won't wound a man, like Hitchens will, with a wit sharpened by many long years of use in fiercely opinionated journalism. And neither will he commend another man to the fire, like Pat Robertson. But the details of his biography do reveal the contours of a strong spine.

By the time Newberg was readying to start his graduate studies at the University of Pennsylvania School of Medicine, in 1993, he wanted to study his own holy trinity: consciousness, the brain, and spiritual experience. But like many people with a scientific interest in spiritual matters, he appeared headed for disappointment. There wasn't much, if any, existing research for him to use as a foundation. But there was one man he had become aware of, the since-deceased Dr. Eugene d'Aquili, who had conducted some research on ritual practices and the brain. Newberg asked d'Aquili for a meeting, and a lunch was arranged.

Newberg was in his mid-twenties, just starting his career. D'Aquili, a doctor at Penn, was roughly twice his age and entirely disinterested. "It was obvious he had taken this lunch with me to politely brush me off," said Newberg. "He didn't have any interest in working with me. And why would he? I had just shown up out of nowhere."

But Newberg didn't accept this polite no—and d'Aquili, seemingly just to get up and get away from the kid at the table, asked that Newberg look over some papers he had written. "Read those," he said. "Then tell me if you're still interested."

The papers were tough, tangled thickets. Newberg read every word. D'Aquili struck him as a creative researcher and a formidable thinker; and when d'Aquili found out that this kid got through the papers and absorbed them, he felt similarly predisposed toward his new protégé. The pair worked together till d'Aquili's death in 1998, and d'Aquili helped design and conduct Newberg's early studies on Tibetan meditators and Franciscan nuns.

For Newberg, the game was on.

"The connection between religious faith and mental disorder is, from the viewpoint of the tolerant and the 'multicultural' both very obvious and highly unmentionable," writes Christopher Hitchens.

Newberg's whole career was first animated by putting this long-voiced atheistic idea to the test: Would some sort of *mal*function show up that explained the spiritual experience? Or would he somehow get a glimpse of God as a kind of ghost in the machine? In his pursuit of an answer, Newberg has become a curator of human spiritual life—taking snapshots of various transcendent states and hanging them on the walls.

For his class, he played some video from one of his most controversial studies, in which he conducted brain scans of a woman awash in the ecstasy of speaking in tongues. Known more formally as *glossolalia*, speaking in tongues is one of the world's most heavily derided mystical practices: skeptics snort at it, and even most religious groups find the act of speaking in tongues aberrant at best, abhorrent at worst.

Among believers, the speaker is said to give himself over to the Holy

Spirit, which takes control of his voice. Practitioners shout and whoop till sputtering flumes of syllables finally come pouring from their mouths. And this noise, they say, is the tongue of angels. Most famously, the R&B singer Al Green is a Pentecostal minister, where tongues remain a staple of spiritual life. When a filmmaker produced a documentary about Green's religious conversion, he withheld all images of Green speaking in tongues until the very end. But the film reaches its emotional climax with a clip of the singer, eyes closed, spouting a torrent of random syllables into the air.

Newberg's video is even more intimate than that. The woman stands before a home video camera, rocking back and forth, her words coming slowly at first, as no doubt she chooses them, until she starts hitting on multisyllabic riffs that sound more spontaneous. Then, finally, she reaches a vibrant, stunning peak—the syllables gushing out of her like water from a hydrant, forceful and uncontainable, moans and shouts. An ecstasy of sound. The kids in Newberg's class settled and resettled in their seats, some visibly wincing as this lady on the screen exulted. I felt a bit uncomfortable myself, in mixed company amid the youth of America, because it felt a bit like watching religious pornography—the sight of a fully clothed woman bringing herself to a spiritual orgasm. Perhaps sensing this, Newberg cut the video short, while the woman still wailed.

Like other religious activity, glossolalia has long been linked by some to mental illness. But a 1979 study in the *Journal of Abnormal Psychology*, along with a 2003 study in the religious journal *Pastoral Psychology*, found no connection between speaking in tongues and mental illness. Newberg's research turned up something even more remarkable. "I know that can be a little difficult to watch," he told his class. "Because it's so unusual if you've never seen that before. But when we looked at her blood flow, in a SPECT scan, we found that she is describing her *experience* perfectly."

As Newberg explained, the woman admitted that in the early stages of speaking in tongues she herself chooses syllables randomly. But at some point she no longer feels as if she is doing the choosing. The sounds just pour from her mouth, unbidden. "That's what we found looking at her brain," said Newberg. "When she reached this involuntary stage, the parts of the brain that mediate and control speech were inactive."

Of course, some might argue, strenuously, with the woman's spiritual interpretation of the experience. They might even call her crazy. But as Newberg told his class, this is a state the woman chooses to enter, a state she achieves in minutes, that does not come upon her at inappropriate times, "in the grocery store or when her children need her," as do the unwanted voices or visions of the schizophrenic. "I happen to know this particular subject," said Newberg. "And she has a normal life. She is a reliable person. And this practice gives her great joy and provides a lot of meaning."

Newberg's research has led him to publicly endorse an "operational view" of spiritual experience: when a practice seems to *work* for the people engaged in it, and brings no discernible harm to those around them, then that practice seems to be one that society can and should tolerate. My guess is that most people agree with Newberg—"most people" being in the middle. But it also seems to me that Newberg's messages of tolerance for belief, and the need for more scientific exploration, are often either lost in the din or purposefully excised from media accounts of his work. Newberg has, in fact, appeared in two films: one, *What the Bleep Do We Know?*, endorsed an extremely mystical view of the world; while the other, *Religulous*, produced by the comedian Bill Maher, endorsed atheism. In each case, Newberg himself appears almost as a different person, a product of each film's editing.

"I am glad to have been in *Religulous*," Newberg told his class. "It was fun to do. But they took out everything I said that expressed any compassion or support for religious belief, and made it look like I had said all spiritual experience was just this trick of the brain."

Though it is disappointing, it isn't surprising that people want to interpret Newberg's work as they see fit. It is the finding of neuroscience, in fact, that belief is at least in part a matter of emotion. Whatever we believe to be true lights up areas of our brain responsible for self-identification and the processing of feelings and sentiments. If we believe something, then, the object of our belief becomes an emotionally potent aspect of our own self-image. There is some common sense to this, too: the most passionate of believers and the most strident of New Atheists are palpably, visibly fired up and ready to defend their positions. And so it follows that nonbelievers might self-identify with the statement "God is a myth," while believers will find themselves reflected in the statement "God is real."

This emotion, this *self*-identification, rather than our faculty for logical reasoning, is why so many interpret Newberg's agnostic data as confirmation of their own worldviews. Emotion also explains why so many engage themselves in the tit-for-tat debates between devout believers and no less committed atheists. Dawkins, Dennett, Harris, and Hitchens all mine the lessons of history to attribute a vast body count to religion; and in retort, unsurprisingly, the Christian author Dinesh D'Souza, among others, notes the number of people killed by secular or atheist regimes like those of Mao and Stalin.

With emotion siphoned from the debate, it seems no group, secular or religious, can claim supremacy on morality—and that both sides often mistake correlation for causation. Consider the suicide bomber, often cited as evidence of the inherent destructiveness of religion in general, and Islam in particular. We tend to associate suicide bombings with religious fanaticism. But ironically, in this, not even correlation is truly present. In his book *Dying to Win*, Robert A. Pape drew on a database of 384 suicide bombers, with known religious or ideological affiliations, who acted between the early 1980s, when the practice was first adopted, and 2003. He found that 57 percent of the bombers represented secular groups. In fact, the Tamil Tigers first perfected the tactic of suicide bombing in Sri Lanka—and they are a Marxist, secular group. Pape further released *Cutting the Fuse* in 2010, which draws on a far larger database of suicide attackers and further extends and supports his thesis that issues of nationalism and foreign occupation are the real motivation for suicide bombing—not religious belief. Scott Atran, an anthropologist at the University of Michigan, has studied Islamic suicide bombers in Palestine and Kashmir, among other places, and found that the reasons they engage in suicide bombing are independent of religious belief—that belief itself is neutral and only channeled, along with more important desires for companionship and self-esteem, into a violent act.

Atran himself happens to be an avowed atheist. But according to the New Atheists, he just isn't critical enough of organized religion. Engaging in an ongoing debate with his colleagues, he fought back and admonished the most outspoken of his fellow atheists for having left science behind: "[They] ignored the vast body of empirical data and analysis of terrorism—a phenomenon they presented as a natural outgrowth of religion," he writes. "The avowedly certain but uncritical arguments they made about the moral power of science and the moral bankruptcy of religion involved no science at all. Some good scientists stepped out of their field of expertise, leaving science behind for the unreflective sort of faith-based thinking they railed against. Sadly, in this regard, even good scientists join other people in unreason."

The most damning aspect of Atran's critique is that the New Atheists don't have any data. Can the scientific field of endeavor, which produces people capable of building atomic bombs and chemical weapons, claim moral authority over religion? "The point is not that some scientists do bad things and some religious believers do good things," writes Atran. "The issue is whether or not there are reliable data to support the claim that religion engages more people who do bad than good, whereas science engages more people who do good than bad. One study might compare, say, standards of reason or tolerance or compassion among British scientists versus British clergy. My own intuition has it a wash, but even I wouldn't trust my own intuitions, and neither should you."

What Atran seems to be suggesting is that the common fiber running through humanity's good and evil acts isn't God or Godlessness—it's us. The source of all our behaviors, good and bad, is multifaceted and rooted in what it means to be human.

Viewed from this perspective, the ongoing debate between believers and unbelievers seems counterproductive—two groups letting their emotions get the best of them. And Newberg's attempt to reconcile science and religion in the field of neurotheology suddenly seems all the more poignant—a way, finally, of uniting humanity's two most dramatic attempts to understand and ameliorate the human condition: the cool rationalism of science and the ecstatic experience of spirituality.

IIIIIIIIIII

"YOU GUYS ALL RIGHT?" Newberg asked. "Are there any questions?"

Newberg asked his class how they were doing, a lot. And he wasn't just talking about class assignments and homework. Sometime in the middle of the semester, in fact, I realized he wasn't only, or even primarily, teaching his students about God and the brain. He was teaching his students how to get along with people who hold different beliefs, teaching them how to live more fulfilling lives, and teaching them to understand and appreciate the limits of human perception and cognition. In his class, "You guys all right?" was a question with philosophical, neurological, theological, and psychological undertones.

Newberg's students had by now been loaded up with information on the workings of their brains. "All my life I have meditated and wondered," one Indian student announced, "why does meditation make me feel so peaceful? Now I know."

Newberg asked if this learning had any impact on the student's thoughts about religion.

"No," he replied, serenely.

"That's good," said Newberg, then caught himself, suddenly. "But it would also be fine if you were changing your opinions," he added.

Newberg took such pains to be respectful that sometimes, like right then, he and his students burst out laughing. His continual assurances, his commitment to respecting every point of view, took a lot of work. "The point is," he continued, chuckling, "you don't have to give up one to have the other. Science and religion can be compatible. They're different ways of looking at, and different ways of trying to understand, the world around us."

What Newberg stressed to his class, however, is that certainty in any of these matters can be very difficult to come by. The religious, no doubt, should accept this—ascribing their belief in a higher power to faith. But even our most basic perceptual and cognitive operations are subject to doubt. Some of Newberg's most entertaining talks, in fact, revolved around the limitations of our ability to understand the world. "The brain receives literally millions of bits of information every second, from all our senses," Newberg told his class. "But we can only be consciously aware of a few bits, a really small proportion, of any of that information, at a given time."

As a result, the brain isn't built to give us a true and accurate perception of reality. There is just too much stimuli to assess; so instead, the brain is built, in an evolutionary sense, to create a model of the world that will allow us to survive. The brain takes processing shortcuts, Newberg explained, bringing the features it deems most important into our consciousness and suppressing the rest. But sometimes the image it creates is wrong. To illustrate this, Newberg showed the class a whole series of perceptual illusions the brain creates. And in seconds, his class was *oohing* and *aahing* at its newfound fallibility.

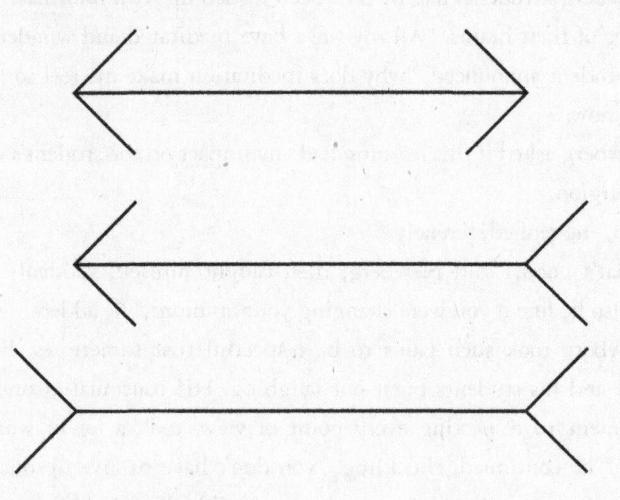

Lines that seem to be different lengths turn out to stretch the same exact distance.

A series of black squares, for instance, separated by bands of white, can create the illusion of flashing gray circles.

Then there is the matter of change blindness: Newberg used a different example with his class, but my favorite is a video in which two teams throw a basketball back and forth for a minute. The first time I watched, I wondered what the point had been. Then some text appeared, informing me that my mind had naturally become so attuned to tracking the dynamic move-

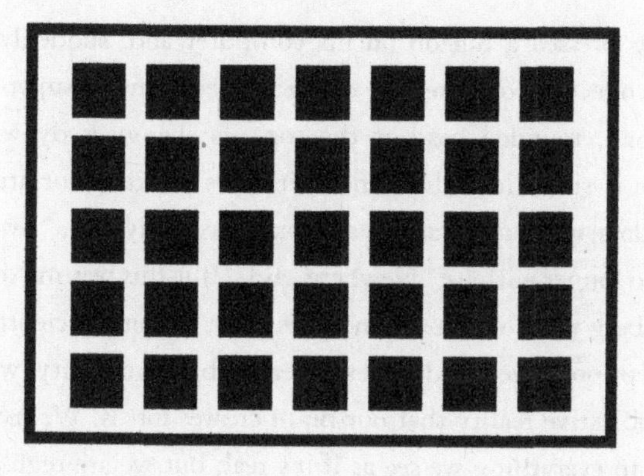

ment of the ball that I didn't see a man in a gorilla suit who moonwalked right across the center of the screen.

Newberg's lesson was simple: sometimes we miss what is right in front of our faces. But then he drove the class even deeper down the byways of philosophy: if our brains are prone to these kinds of errors, if we are only consciously aware of a small range of stimuli, how do we know our perceptions are accurate? "Well," Newberg said, answering his own question, "we *don't*, in any definitive sense."

All our visual perceptions are created by light reaching the retina of our eyes, which is then translated into a neural signal and interpreted by the brain. To demonstrate this for his students, Newberg showed them a series of three slides, each depicting what he called the "same familiar object."

The first image was all lines and shades with no discernible structure at all. "This is the raw pattern of light our retina receives and sends to the brain," said Newberg. "This is what we *actually* see, and our brain manufactures it into an image."

Newberg tapped his laptop and brought a second image on to the projector screen. "I'm not going to show you all the stages of processing," he said, "but this is about halfway there."

In this second frame, the image has begun to cohere. Now a foreground is distinguishable. And there is a figure coming into focus at the center—small and rounded at the top, flaring out at the bottom. But what was it? No one could tell. "Okay," he said. "Now we're going to jump ahead, all the way to the end."

Newberg pressed a button on his computer and, suddenly, out of the amorphous collection of pixels, a dog emerged. And a super-cute one at that—its small, rounded head at the top, its shaggy body widening out beneath it as it stares into the camera, its eyes pleading for attention. The girls in the class, and I confess, this author, collectively said, "Awww."

"He is no longer with us," Newberg said. "But this was my dog, Cosmo."

As Newberg went on to explain to his class, cognitive scientists studying human perception agree: we don't experience objective reality; we experience a model of objective reality that our brain creates for us. We have no choice but to react to everything we see as if it's real; but we are regularly missing all kinds of information our brain deems unimportant. This may not seem like that big a deal until we consider all the errors of our perception and the philosophical problems that come with this state of affairs. Just what was Cosmo? A cute dog? Or the formless hoo ha Newberg showed us in that first image—a raw pattern of light striking matter?

Philosophers have long been caught up in a debate over whether or not the world we perceive is essentially the world that is so. But for our purposes, the important thing to understand is that we also endure these same kinds of perceptual challenges when processing our own thoughts.

In the introduction, I gave a brief primer on the role the amygdala plays in our cognition. And Newberg also told his class about this perhaps most problematic part of the brain. These almond-shaped structures at the base of our temporal lobes are complicated, playing some role in mediating all the input we receive—from the interpretation of facial expressions to the processing of emotions and memory recall. The amygdala's chief job is a simple one. It is primarily our own personal bodyguard, shouting danger any time it perceives a threat. In this respect, it is much like all our tools of perception—chiefly concerned with helping us to survive. The problem is that it is activated by both bodily danger and threats to our worldviews. It makes difficult conversations that much more difficult, causing us to feel anxiety and nervousness and the tense emotions associated with the fight-or-flight response.

We normally think of fight or flight occurring when we are confronted with some physical danger, like a man with a gun. Do we run, or try to take the dude out? But in the human brain, a contentious conversation between believer and atheist, or Republican and Democrat, provokes similarly stressful biological and neurological responses—including anxiety, sweating, increased heart rate, and the sensation of anger or even fear.

The amygdala is so closely associated with our fight to preserve our lives that the amygdalas of cancer patients undergo changes in size. And of course, this suggests a negative feedback loop in which stress and fear lead to greater stress and fear. Underscoring this, terror management theory (TMT), a new field of psychological study, has equated perceived threats to our beliefs and the threat of death itself. Reminders of our mortality, and of course such reminders are all around, render us more sensitive to and guarded against threats to our beliefs. The findings have held true for both atheists and believers. In fact, while we never managed to schedule an interview, one of the masterminds of TMT, Tom Pyszczynski, did tell me in an email that the popular focus on the religious as the only ones denying their own mortality is misguided: "Science, evolution, atheism, etc., can all take on a 'religious' function in that it becomes a faith beyond question," he writes, "and functions to manage fear, especially fear of death. There's a lot of signs of this happening in the world today."

No doubt, then, this unruly and illogical structure of the brain affects both sides in the ongoing debate between atheists, who think of religion as threatening to the reason they hold so dear, and believers, who think of religion as providing the source and foundation of all life. But what Newberg has found, by bringing science and religion together, is nothing less than revolutionary: because mystics of varying religious persuasions may or may not have found one true God; they may or may not have discovered the identity of Cosmo the dog; but they most definitely have found a way to render our emotional perceptions more accurate. They most definitely have found a means of settling our unquiet minds.

||||||||||||

IT IS EARLY DECEMBER 2009, and the lights at the Church of the Epiphany in Oak Park, California, are turned down low. Two men are on stage. One is Yuval Ron, a Grammy- and Oscar-winning musician and peace activist. He is dressed all in white. His close-cropped brown hair is lost a little in the shadows on stage, where he hunches contemplatively over an oud, an ancient, pear-shaped, short-necked member of the lute family. The music Ron plays is slow and repetitive—traits not normally associated with entertaining music. But the rhythms and melodies he picks out tonight fill and structure the room, their repetitive nature creating a stable space.

Periodically, the other man steps up to a microphone at stage right. Mark Robert Waldman is tall and skinny, and here in this Christian church he wears a Muslim hat—or *kufi*. He speaks in sober bass tones, and what he says captures the jargon of science and the self-help industry: "Music, spirituality and the brain," Waldman intones. "*What*, in heaven's name, do these words have in common? As you contemplate your deepest values and beliefs you can change the structure and function of your brain in life enhancing ways. The right meditations can reduce anxiety and depression; can lessen symptoms of illness, and improve the cognitive function of your brain; create inner peace and enhance the neurological capacity for compassion; and maybe, just maybe, slow down the aging process of the brain."

Over the course of the night, between stretches of Ron's meditative music, Waldman dispenses advice: If anyone in the audience wants peace and happiness, they can get it through meditation and prayer. If music is played in the background, it should be pleasant, slow, repetitive and relaxed. And the prayer or meditation itself should be profoundly meaningful to the practitioner and positive in all its connotations.

This night isn't a scientific presentation. So Waldman makes only passing references to the neurological findings that support his claims. But he promotes various mystical practices—including meditation and the Christian centering prayer. And in perhaps the most dramatic moment of his performance, he raises his voice like a priest to wring some emotion out of a poem: "I am circling around God," he says, "around the ancient tower, and I have been circling for a thousand years, and I still do not know if I am a falcon, a storm, or a great song."

What I never did hear him tell the audience, however, is pretty remarkable for a man dressed in a kufi and reading mystic poetry in a Christian church: Waldman sees God only as a useful and beautiful metaphor—not a scientific reality. He doesn't believe in any religion. But he is a nonbeliever with a newfound respect for religion, a respect so deep he can't always be distinguished from a believer, because he has come to accept the positive power of some religious practice.

A counselor, teacher, and writer working in Camarillo, California, Waldman was content enough in his own non-belief that he felt religion held little value for psychological or societal health. But when he began to co-write and collaborate with Andrew Newberg, his viewpoints radically changed. And their partnership quickly went much further than either of them expected. A couple of books later, in fact, Waldman is firmly established as Newberg's sidekick.

Newberg conducts the lab research. Waldman combs the library stacks, integrating related scientific papers into their books. Working in this manner, they soon found themselves on the fuzzy borderland between science and religion, on ground the two could share. And from that position, they found they weren't just identifying the neural correlates of religious experience, they were finding that specific spiritual practices *work*.

Consider Newberg's early research on Tibetan meditators and Franciscan nuns, who Waldman referenced to the church crowd when he mentioned the centering prayer.

Used by Christian mystics since the fourth century, the centering prayer grew in popularity in the 1970s and differs from standard meditation in one crucial respect. In meditation, the practitioner sits and observes the workings of his or her own mind, no longer reacting to the flurry of thoughts constantly generated by the brain. In this state, all thoughts are allowed to just sail by, without any emotion or judgment. The meditator might choose to remain in this condition, allowing her sense of awareness to blossom; or she could shift focus to her breathing, to a particular ideal, or to a mantra. For the Christian mystic, however, engaged in a centering prayer, the object of focus is specific: communion with God.

The correlation between the goals of these two mystic practices and the neural processes Newberg observed is total: the Tibetan meditators Newberg studied felt incredible peace; correspondingly, activity in their amygdala and their rational frontal lobes quieted. They also claimed to feel a oneness with all things; and indeed, the areas of the brain governing the position of their bodies in relation to other objects fell silent, too. The nuns experienced all this, plus a sense of communion with a higher power; and upon analysis, their brain scans looked much like those of the Tibetans, with the addition of significant activity in the limbic or emotional center of the brain, reflecting their profound sense of union with God.

If that was all there is, the skeptic might feel justified in snorting. To them, this would undoubtedly seem like Newberg had simply found the neural correlates of delusion. In retort, I paraphrase a pop song: *How can something so wrong, feel so right?* Because, scientifically speaking, the experience *is* right. The sensations these practitioners report aren't delusion; they are the self-directed workings of the human mind, like a horse put under harness. And even more important, these positive changes in brain function, if practiced enough, transform our baseline mental states in incredibly healthy ways.

The amount of scientific research into the neurological effects of prayer and meditation is still small, but it is growing quickly. And what we're finding is that short-term changes in our consciousness, during contemplative practice, produce long-term, positive neurological effects. People tend to think of their personalities and ways of being as somehow fixed. And in science, these traits and flaws alike have been linked to brain function. But as Waldman put it to me, "The whole notion that our brains are hardwired for much of anything is wrong. The name of the game is neuroplasticity."

The concept of neuroplasticity is now an accepted fact of neuroscience: our brains literally rewire themselves, creating new connections among neurons based on what we do and think. During the 2010 Winter Olympics, television cameras often captured skiers in mid-reverie: waiting for their next run, poles in hand and eyes closed, twitching and turning their heads side to side as if slaloming downhill. The skiers were meditating on their craft, and with good reason: a brain that carries out a specific set of calculations and functions, again and again, even in our imagination, literally reshapes itself to the task. That's neuroplasticity. And that is precisely what's so powerful about a regular meditative practice.

In the same way these skiers imagine their next downhill runs, everyone can watch their negative thoughts float by without judging them or reacting to them. As a result, the next time anxiety threatens to derail their efforts, they're more likely to let go of all that fear and stress and keep working toward their goals. Need to have a difficult conversation with the boss or your spouse? Imagine it, repetitively. Notice all the anxiety and fear associated with the conversation, but don't react to it. Instead, imagine yourself simply saying what you need to say. By the time you're actually engaged in the conversation, you'll feel like you've pulled it off successfully numerous times before, and you'll better understand how to function, despite whatever stress you still feel.

Meditation and prayer have for too long, culturally speaking, been relegated to the margins. But opportunities for meditation are all around us. We may engage in them without realizing it. In the bestselling book *Born to Run*, in fact, longtime men's magazine writer and running enthusiast Christopher McDougall slowly relates the practice of jogging to meditation. Think of a disciplined runner maintaining focus on the ground as it moves beneath her feet. The runner does not gaze out toward the horizon or allow thoughts of the remaining distance to hinder her. This runner is meditating. And today's ultramarathoners, who run distances of a hundred miles, are essentially engaging in what the Buddhists called "running meditations."

A scientist we met earlier in this book, Dean Radin, meditates during his workday by staying strictly focused on the task at hand. While most people think about the next, most dreaded items on the to-do list, Radin stays in the present, practicing a classic mindfulness meditation. Do this sort of thing often, do this every day, according to Newberg and Waldman's findings, and the benefits will change your brain function—and by extension, the quality of your life.

Remember the meditators Newberg began his career studying? Their brains appear to be shaped over many years and long practice into an enviable configuration—a state of relaxed concentration and lower anxiety. In short, those dudes in the saffron robes seem so peaceful because their brains *are* more peaceful—their experience of life is more peaceful.

But don't let their expertise intimidate: we can get great benefits from meditation or prayerful practice almost immediately, in a variety of areas—from skiing, to creativity, to spirituality, to compassion, to our faulty memories.

Newberg conducted an experiment in which subjects improved their memories by meditating just twelve minutes a day, for only a few weeks. And as Waldman told me, to increase your own sense of well-being, "You don't have to believe in God. You can just think about what you most value in life—compassion, reason, love, peace—and you'll literally strengthen the neural connections that enable you to carry those values into the rest of your life."

Yet this promising practice also bears a threat: because if we quite literally are what we think, we are better off focusing on positive ideas. In short, if you choose to worship a God, science is now capable of telling you what kind of God is most worthy of that worship—what kind of God will bring you the greatest sense of peace and best enable you to relate to your fellow man. In this vein, Newberg and Waldman's research suggests that people who focus on a judgmental, vengeful God probably have less healthy brains—brains filled with fear and anxiety. Better not to take your marching orders from Pat Robertson and Jerry Falwell or, less famously, the eight-hundred-plus-member Rabbinical Alliance of America, who link terrorist attacks or natural disasters to God's wrath at homosexuality; better to focus on ideas like love, unity, and peace.

Atheists, too, should consider: The more evil they attribute to religion, the more that view will become their reality, and the less able they will be to rationally consider contrary evidence.

Waldman has become something of an evangelist for the power of meditation. He sees his stumping for contemplative practice as promoting science and, at least, a broadly defined spirituality with a body of research that is growing quickly. Scientists working at hospitals and universities in Massachusetts and Minnesota, associated with Harvard and the Beth Israel Deaconess Medical Center, have demonstrated that meditators practicing relaxation techniques don't just change their brain function. They even undergo genetic changes. In both experienced and novice practitioners, the genes affected were related to a pronounced *lack* or reduction of stress. The gene expression of beginning meditators, in fact, changed in just eight weeks of regular meditative practice.

This is not so crazy as it sounds. Our genetics, after all, represent only potentials. Genes that are activated by environment, or our actions, are said to be "switched on" or "expressed." As an example, research published in 2007 in the online journal *BMC Genomics* showed that smoking leads to permanent changes in gene expression. Genes that are irreversibly changed may help to explain why former smokers remain more likely to develop lung cancer. The Benson-Lieberman study on meditation, however, is particularly dramatic, demonstrating that positive molecular shifts can occur in as little as two months, suggesting even beginning meditation practitioners are less likely to incur the health problems associated with stress.

With findings like these, momentum is building: some may balk at Eckhart Tolle's particular brand of mysticism, but his sales figures speak to a possible shift in the culture, away from being indoctrinated into a religion or lectured to about spirituality and toward experiencing it firsthand.

Most profitably, from a scientific perspective, Dr. Susan Smalley, a psychiatrist and behavior geneticist, took up meditation after a cancer diagnosis and founded the Mindful Awareness Research Center at UCLA. She is documenting meditation's ability to lower blood pressure and boost the immune system; increase attention and focus, even for those suffering from ADHD; help with difficult mental states, such as anxiety and depression; increase a personal sense of well-being and make practitioners less emotionally reactive; and strengthen the neural networks responsible for decision making, emotional flexibility, and empathy.

So, if meditation is so great, why aren't we all already doing it?

One reason is that contemplative practices require work.

Newberg is in fact concerned that too few religious believers engage in the dedicated prayer and meditation he advocates. The time and effort necessary can appear too much for busy people with busy lives. Further, meditation is associated with Eastern religion, and not all Christian faiths have an equivalent practice, like centering prayer. Worse, persuading strict materialists, atheists, and agnostics to sit down on a prayer mat might be even harder. Because anything associated with religious faith is something they are likely to reject, out of hand.

A rational person, acting purely on reason, might figure that Tibetan Buddhists have practiced meditation for millennia because *it works*. But among the New Atheists, only Sam Harris has been a proponent of meditation and contemplative practice.

In short, dedication to a worldview can cause even scientists to behave irrationally. As recently as 2005, in fact, a coalition of neuroscientists protested a planned speech by the Dalai Lama at a neurological conference. The online petition collected roughly a thousand signatures. "I don't think it's appropriate to have a prominent religious leader at a scientific event," one of the petition's organizers told Britain's *Guardian* newspaper at the time.

In many respects, this little scientific tempest mirrors our current culture war. Consider what happened even more recently to Dr. Francis Collins, a world-class geneticist and administrative head of the human genome project, which mapped the structure of human DNA. Collins also happens to be an evangelical Christian. And evidently, that was his mistake. I contacted Collins in his capacity as founder of the BioLogos Foundation, a group devoted to demonstrating how faith is compatible with evolution. But after he was selected by the Obama administration to serve as director of the National Institute of Health, a spokesperson at his foundation told me he couldn't talk. The controversy over his faith was too great for him to stick his poor religious head above ground. Some particularly adamant New Atheists were publicly claiming his beliefs undermine his ability to practice science.

The objection of some to Collins's appointment, or an appearance by the Dalai Lama, strikes me as particularly telling. Collins's book, *The Language of God*, calls on Christians to embrace evolution. And the Dalai Lama openly encourages his followers to embrace scientific findings: "As in science," he said, "so in Buddhism." But for some, the perceived threat that Collins or the Dalai Lama presents is just too great: *Science and religion, compatible, collaborative, or even civil? No. They just can't be!*

But what Newberg and Waldman are demonstrating is that the practice of science and the pursuit of spiritual life *are* compatible. Newberg and Waldman are pointing us toward an area of life—prayer and meditation—in which the scientific method validates spiritual practice. And this should have ramifications on how we all view the debate between atheists and believers.

Human progress is marked by shifts in our thinking and philosophy. Purely magical thinking, for instance, in which causal effects are not understood at all, is slowly replaced by scientific reasoning. And as one system of thought displaces another, many people tend to reject all that came before. No doubt, in American intellectual life, we're struggling with that stage right now. The rise of Dawkins, Dennett, and Hitchens represents just that kind of dogmatic adherence to a particular way of looking at the world. But over time, as described in the writings of the philosophers Jean Gebser and more recently Ken Wilber, an enlightened society begins readmitting the truths revealed by earlier methods. And I think, in Newberg and Waldman's books, we're seeing just that kind of inclusiveness. We're seeing the foundation for a worldview in which science and spirituality serve each other.

This is one story, in fact, in which all the narratives and subnarratives end well. Collins has taken his post. And the leadership of the Society for Neuroscience, which invited the Dalai Lama to speak, did not back down. In his address, the Dalai Lama echoed the findings Waldman shared with his audience in California—that there are ideas we each carry in our heads and hearts, ideas of compassion, tolerance, caring, and consideration, that "transcend the barriers between religious believers and nonbelievers, and followers of this religion or that religion."

It has now been scientifically demonstrated that if we focus on these kinds of ideals, we have the power to change ourselves at a genetic level. We have the power to change our brain structures and our ways of being—peaceful or angry, compassionate or nasty, fearful or confident, stressed or relaxed—from the inside out. And the lesson for us here, that we can take firmer control of our own minds, is something mystics have known and carried out for millennia, long before science had the means to investigate it.

For those who think science is the only way of knowing, who think that until something is validated by science it might as well not exist, this long time lag between the mystics' discovery and the ability of science to corroborate it, might be a bit humbling. And of course, for anyone who believes that anything arising from religion is necessarily tainted, there must surely be some urge now to argue the efficacy of prayer and meditation away. But for people who are willing to look at both science and religion for whatever truth we might find there, for whatever treasure might be ours, the findings of Andrew Newberg are a stepping stone to a better, happier, more fulfilled life. And to see that in action, we need look no further than Newberg's research partner, Mark Robert Waldman, the nonbeliever who found himself spending a shocking amount of time in churches, synagogues, and temples.

"I never saw this coming," Waldman told me. "I thought I had things pretty well figured out. But my work with Andy has changed my life, and much for the better. Because of my meditative practice, I'm more patient and compassionate and enjoy my relationships and my work more. I enjoy the time I spend in these different places of worship immensely. And I plan on continuing this work and spreading the word."

|||||||||||

NEWBERG HAD ONE LAST trick up his sleeve—a lecture in which he promised his freshman class he would try to "tie all this material together."

He had a lot of ground to cover.

In more than ten years as a neurotheologist, just what had he learned? And where had he taken his class?

Well, he said, we now know that all our experience takes place inside our heads, and so all experience is subjective. The power of science is that we can use it to corroborate subjective experience, to demonstrate the validity of our perceptions. But measuring instruments change, and science changes, and reality seems to shift with our ability to know it. The idea that microscopic creatures caused illness seemed absolute lunacy—until the microscope was invented. Post-Newton, it seemed we live in a universe that looked a lot like a billiard table, in which everything moved according to how it got smacked. Then technology improved to the point that we could perceive another system of operations beneath the Newtonian realm. This new world of physics was quantum mechanics, and we're still reflecting on what to make of that.

This doesn't mean we should believe any idea that comes along, Newberg said, or any idea we just happen to like. But it does mean we should live in the awareness that the ground may shift under our feet. And it does mean we should admit to the range of possibilities reality might hold. Both scientific and spiritual. It is just that, neurologically, we're not built to live with that kind of ambiguity. "The brain has a propensity," said Newberg, "to dismiss ideas that conflict with the way we see the world. If someone disagrees with us, our brain starts sending anxiety messages. Because our brain wants to preserve this view of the world it's constructed, and in order to do that, the easiest thing to do is reject this person."

So, what happens in a disagreement?

According to Newberg, we might explain our position rationally at first. But if our reasoned argument doesn't persuade our opponent, it isn't our instinct to say, *Well, they just have a different way of looking at the world.* "Our instinct," says Newberg, "is to say, 'This other person is irrational!' "

Newberg laughed. So did the class.

"Right?" he said. "We've all been there. 'You don't agree with me? You must be crazy!' "

Neurologically speaking, the entire debate between the most extreme of believers and unbelievers boils down to this—both sides talking past the other. So what has Newberg learned of religion, of spirituality? He's learned that modern scientific instruments corroborate the experiential validity of some of the most outrageous perceptions man has ever claimed: the feeling of divine union, the sense of unity among all things. But are these subjective experiences pointing to some objective truth about reality?

Or, as Newberg put it, "Is it possible, that when we turn off certain parts of our brain, or pay less attention to them, we're better able to access the input we're receiving from other areas of the brain? And is that information *real*, or just the result of focusing our awareness?"

When you stop being concerned with yourself, with your hopes and fears and your iPod and the chair you're sitting on, does other, real input you're receiving and usually unaware of come to the fore?

Newberg is a scientist and is humble enough to admit he doesn't know the answer. And as he points out, from a definitive scientific or epistemological perspective, no one credibly can. "But as a scientist," he said, "there is something really interesting to me about these mystical states. Because when you ask people about these spiritual experiences, they don't describe them in the same terms we describe dreams or hallucinations."

In the case of figments of imagination, once people are back in everyday reality, they recognize the dream as having been less real than, for instance, the bed they sleep on. "But when someone has a mystical experience," Newberg said, "they still look back at it and say it felt real. In fact, they claim the mystical reality felt *more* real than the one we're in right now. And all of it points back to this sense of unity—of the self dropping away, and of feeling connected to something infinitely larger."

There is irony in this.

As Newberg told his class, science tells us something much the same: science tells us that everything arises from the same source; everything emerges from the same small spot of the Big Bang; everything is comprised of the same whirling particles; the same waves of energy; and at these fundamental realms of reality, there is this capacity for separate particles to be connected. "There is obviously more work to do," said Newberg. "But isn't it funny? There is this long fight between science and spirituality. But if we really pay attention to science and to profound mystical experiences, they are both essentially telling us the same thing—that this profound connection exists, lying underneath it all."

Newberg ran through all this in about an hour, hitting this last point just before he ran out of time.

There was a sudden silence then, as everyone realized this was the end.

His students, smiles beaming, broke into an enthusiastic round of applause.

Newberg, smiling sheepishly, bowed.

And that was it.

Class dismissed.

The Impossible Dream

Sleep as an Untapped Resource for an Awakened Life

All men dream: but not equally. Those who dream by night in the dusty recesses of their minds wake in the day to find that it was vanity: but the dreamers of the day are dangerous men, for they may act their dream with open eyes, to make it possible.

—T. E. Lawrence

I trudged up a steep set of steps, toward a large, open meeting space on the second floor of what was essentially a massive wooden hut. I was running late. And the buzz of conversation was already going on above me.

I was here, having just arrived in Hawaii, to explore the world of dreams, to attend a workshop about a very specific kind of dreaming. And the introductory session was about to start. As I hit the top stair, I saw this was going to be an informal gathering—a cross between an academic lecture series and a hippie commune. Metal folding chairs were arranged in a small semi-circle, and the walls featured big, screened windows, to permit the cool evening air inside. The interior ring featured a series of small mats for people who preferred to sit on the floor. And the director of the workshop, Stanford University sleep and dream researcher Dr. Stephen LaBerge, was already present, barefoot, and dressed in shorts and a floral print shirt.

LaBerge made his way around the room, shaking hands with all thirty-two attendees. He was slim, sported a stylish gray haircut, and introduced himself simply as "Stephen."

After all these years, it was a pleasure to meet him up close. I had first run across his book, *Lucid Dreaming*, about sixteen years earlier, in college. And I still vividly remember dismissing it out of hand, after just a few dozen pages, because it seemed too good to be true.

Lucid dreaming is that cool. But what is lucid dreaming?

Well, the key difference between a typical dream and a lucid dream is simply one of awareness. An average dream occurs like a David Lynch movie, in evocative fragments. We experience the event, intimately, feeling terror, anxiety, or lust, whatever the dream provokes. But we remain locked in the idea that everything we see is externally real. We remain unable to consciously choose our actions with the knowledge that all this is . . . *but a dream.*

The lucid dream stands this entire dynamic on its head. In a lucid dream, we gain awareness of our true state. And lucidity strikes like lightning, rendering everything in the dream—every wall, every person, every color, every speck of light—in detail that often transcends waking perception. But even more important, we are now truly *in* the movie—and we can choose our actions accordingly.

Want to fly?

Take off! Like Superman, or Neo. The laws of physics play no part here. *Want to have sex with that dream hottie?* Go for it. It's a dream. That's one fantasy figure who will most likely say yes.

A lot of popular media accounts focus on these sensational aspects of lucid dreaming. And here in the United States, lucid dreaming is best known from movies like *Inception* and *Vanilla Sky*, both of which take full advantage of the anything-goes atmosphere of the dream. But popular depictions usually miss the larger, more fundamental picture: lucid dreaming is intimately related to the practice of meditation; and for many practitioners, the two pursuits are really one. It is a phenomenon that most definitely exists, but because it has connections to religious practice, raises important questions about the murky workings of consciousness, and takes seriously the irrational world of dreaming, the lucid dream carries the whiff of hoo ha.

LaBerge addressed the first link, between meditation and lucid dreaming, in his opening talk. He took his seat at the front of the room and began arching his thick eyebrows in various configurations, like a Mad Hatter, until we all fell silent. Then he spoke: "Dreaming and waking," he hollered, his face turned toward the ceiling. "Waking and dreaming . . ."

"We're awake right now," he said, arching his eyebrows again. "Or, at least, it *seems* so, doesn't it?"

LaBerge let this observation hang in the air for a few seconds. "But something you should consider," he said, "is that we so rarely question what state we're in."

Over the course of the next week, LaBerge promised, he would teach us the differences and the similarities between waking and dreaming. He also advised us to look at the pursuit of lucid dreaming as a metaphor for our entire lives. "Consider," he said, "how often are you aware of your surroundings, *really* aware? And how often are you merely *reacting* in the same automatic way as you do in dreams?"

Given this question, the connection between meditation and lucid dreaming appeared obvious to me. In fact, after conducting my research into Andrew Newberg, I had pursued a meditative practice of my own. I had also, in preparing for this trip, experienced a particularly dramatic lucid dream. My first. But I figured I got it: through each practice, the doer gains greater awareness and ability to choose. *Who am I going to be right now? How would I like to respond?* Meditation can help us lead more fulfilling waking lives. Lucid dreaming can grant us awareness even in our sleep. But starting the next morning, I was about to learn how thin the line between waking and dreaming really is—and how easily I could improve my experience of life, just by more closely observing the world around me.

IIIIIIIIIII

THE SUN HADN'T BEEN up all that long by the time I strolled across the lawn, looking for coffee. The light was dim. The grass was still wet. And in the distance, I saw a blurry little shape scrambling around, flaunting the agility of a cat and a top-end speed I'd seen only in the mice shooting across the floor of my first Philadelphia apartment.

Because I had arrived at night, this was my first view of the Kalani Oceanside Resort in Hilo, on the big island of Hawaii. The word *resort* suggests—to me, anyway—sleek, corporate spaces and luxury. But Kalani is rustic, an aging compound of communal living spaces and hut-style accommodations for those more interested in privacy. There was a small general store and a tiny snack bar. A wooden deck maybe forty yards long comprised the common dining area, dominating the central space. There were some sections set aside for group yoga practices, and meditation. And that was pretty much it. The ocean sat across a narrow band of highway. All else, in the near distance, was jungle.

I bee-lined to the deck, saw the coffee pots, and felt a wave of relief before I even took my first sip. I was struggling with the time change. As I poured a cup, I spoke to a Kalani staff member who was setting up for breakfast. It was 6:30 AM, Hawaii time. But the girl, ritually tattooed and dreadlocked, her face fresh-scrubbed and clean, looked like she had been awake for hours.

I described the small brown blur I'd seen across the lawn. "What was that?" I asked.

"A mongoose," she said.

I paused a beat, remembering Rikki Tikki Tavi, the mongoose from *Jungle Book*.

"They're Hawaiian squirrels," she explained. "They're everywhere. But they're shy. You see them around, but not for long, because they dart away."

The metaphor, between elusive mongoose and ephemeral dream, was obvious. Though everyone dreams, many people claim not to remember their dreams at all. And how many times had I awakened, over the years, remembering a snatch of what I'd dreamt only to find that the harder I pursued it, the faster it seemed to vanish?

I sat down and stared out across the landscape through the still misty air. This trip to Hawaii marked the final step in a nearly two-year immersion in the paranormal. I had explored near-death experiences, telepathy, and

UFOs, among other taboo topics. But my experience with Andrew Newberg and meditation had been the most inspiring. After that, I felt keen to follow the path into inner space. So keen, in fact, that I felt thwarted when a voice behind me interrupted my thoughts: "Are you Steve?" a man asked.

I turned to look over my shoulder, feeling grumpy. I was already going to sit in a meeting space, shoulder to shoulder with thirty-two strangers, for a few hours each morning and again each night. I was jetlagged, and I hoped that I might at least enjoy some alone time at 6:30 A.M. "Yes," I said, turning to look at the man behind me, my voice tight and clipped. "That's the name."

"I'm Steve, too," he said. "I saw your name badge at last night's session."

This second Steve was gray-haired, balding, and wrapped in a colorful Tibetan robe. He wore sandals and smiled way too brightly and enthusiastically for the hour. *Oh no,* I thought. *Aging hippie*.

I asked him if he wanted to sit down, barely feigning politeness, my mouth curled into a sour smile. Steve sat down at the table next to mine and talked to me from several feet away.

In the next few minutes, I learned that he was from Texas. He had been a practicing Tibetan Buddhist for decades.

In the form of Buddhism he practices, in fact, lucid dreaming has long been pursued under the name "dream yoga." The goal is to become fully aware in both waking life and dreaming life—aware, twenty-four hours a day.

Among some Buddhists, becoming aware during sleep is a way of rehearsing for the transition into death. "The idea," Steve told me, "is that if you don't learn to become aware in your dreams, you won't become conscious in death."

Heavy stuff, but still I felt put out, only half-listening. Then Steve switched gears abruptly.

"You been to the general store yet?" he asked.

"No," I said. "I got in late last night."

"I am hoping I can get a knife," he explained. "They took my knife from me yesterday at the airport, and I have to tell you: A Texan is naked without his knife."

He shook his head, ruefully. "In a place like this . . . ?" he said, motioning around us, toward the wilds of the jungle.

I laughed. And remembered LaBerge's suggestion of the night before: because until this moment, I had just been *reacting* to Steve. I had taken one look at him—old, gray-haired, wide-eyed, sandaled, and be-robed in Eastern garb—and got immediate feedback from my brain, reflecting all of my own biases: *Flake*, I thought. *Hippie*.

It was several minutes after I reluctantly engaged with him, when he *surprised* me, that I became aware of him at all: this was a man invested in the teachings of the Far East but still mindful enough of what he learned in the great West—to *want a knife*—to believe there might be something around here, at a lucid dreaming workshop, that he'd need to stab.

I didn't understand this combination of seemingly incongruous personality traits—and that was the point. Steve had now emerged as a complicated, three-dimensional human being. The clichéd hippie was someone I had only imagined. I *liked* this second Steve and wondered what other surprises he might have in store. And I *got* LaBerge's previous talk now, in a very personal way: How many people, how many situations, pass us by, unexamined, in the course of our waking lifetimes? How much of our waking lives do we spend, essentially, asleep—thinking we know someone or something and *reacting* to that internal judgment, when we don't really know anything at all?

I had done the same thing, many years earlier, when I allowed stupid cynicism to keep me from reading LaBerge's work when I first encountered it. I only overcame my resistance to the concept of lucid dreaming later, when the movie *Vanilla Sky* was released, and the publicity surrounding it included some mentions that—hey, what do you know—this lucid dreaming stuff is real.

LaBerge understands this resistance, which shaped his life and career. But for him, lucid dreaming was always part of reality. Like many people, he had lucid dreams, spontaneously, as a child. In his, he was an underwater pirate, totally untroubled by the lack of oxygen.

He was surprised, years later, to find that mainstream scientists denied the existence of lucid dreaming entirely. Yet science was always his path.

He was raised Catholic. And in grade school, he asked a nun about the Catholic doctrine of Transubstantiation—the teaching that the wafers doled out during communion become the body and blood of Christ. "I was a rationalist," he told us. "Even then. And so if the wafer had become different, it would have to become *literally* different. So I asked her, '*How could this be so?*' "

The nun issued her automatic response. "That," she replied, "is a holy mystery."

To the young LaBerge, this was proof that religious people were just stupid. It was an opinion he didn't hold forever, but an opinion that nonetheless illuminated the road he would take into science.

LaBerge is in his sixties now; and as it happens, he grew into adulthood in *the* '60s. He wanted to examine consciousness scientifically. But back then, consciousness was a dirty word in academic hallways.

Because science can only weigh, measure, and objectively assess that which can be weighed, measured, and objectively assessed, the subjective inner states that make up our lives were, and sometimes still are, relegated to the margins. As a result, the field of psychology was still mostly dominated by behaviorism, a worldview in which inner states and thought processes didn't matter; only our actions, our behaviors, claimed real scientific importance.

Frustrated, LaBerge spent some time in the '60s counterculture, attempting to study consciousness through the use of various illicit substances. But like Catholicism, hallucinogens didn't take. "I quickly understood," he said, "that there must be a better, safer way to go."

For LaBerge, that meant pursuing a doctorate in psychophysiology at Stanford, in 1977, under the mentorship of sleep researcher William Dement.

Psychophysiology was a kind of halfway house for LaBerge, concerning itself with the physiological basis of psychological responses. "It wasn't what I wanted," he told the group in Hawaii. "But at the time, it was the closest I could get."

LaBerge had it in mind to study lucid dreaming. But in scientific terms, this was a nonstarter.

Lucid dreaming has long been a part of human history, yet here in the West it's been relegated to the fringe.

Aristotle had mentioned lucid dreaming. Buddhists had been actively pursuing mastery over the dream state, along with their meditations, for thousands of years, as had the Aborigines, who considered the dream time sacred. Lucid dreaming also turns up in Islam and Hinduism.

A French researcher named the Marquis d'Hervey Saint-Denys wrote the first-ever book about lucid dreaming, in 1867, translated as *Dreams and the Ways to Direct Them: Practical Observations*, which he published anonymously because colleagues sneered at the topic. But it was Dutch psychiatrist Frederik van Eeden who first coined the term "lucid dream."

Van Eeden studied the phenomenon systematically in himself. But in 1913, having completed *The Study of Dreams*, he could only get a hearing from the Society for Psychical Research, an organization known chiefly for studying—what else?—the paranormal. And so by the time LaBerge determined to study the phenomenon, lucid dreaming was still, from the point of view of science, the bearer of much Paranormal Taint—and as real as the tribe of faeries living in Richard Dawkins's garden.

Sleep was considered a state of *un*consciousness, and that was the problem. Lucid dreaming paradoxically required the dreamer to be both conscious and *un*conscious at the same time—an impossibility! But LaBerge had something on the scientists of the day. Experience. To him, the question wasn't whether lucid dreaming was real. The question was whether or not technology and experimentation had caught up to lucid dreaming, whether science had grown enough in its capabilities to catch up to the Aborigines—and begin study of this intriguing state.

LaBerge partnered with researcher Lynn Nagel, hoping to use a finding discovered by their academic director, Dement. In 1962, Dement was studying REM sleep—or rapid eye movement. REM is often described as a consistent left-right motion of the eyes. But LaBerge says REM would be more aptly abbreviated HJM—for "Herky Jerky Movement," because the motion of the eyes is in fact so frantic and random.

This is key, because one night, as Dement observed a research volunteer in REM, he noticed something odd. Suddenly, the sleeper's eye movements shifted. The wild, jerky movements typical of REM suddenly *did* assume a regular rhythm. Left, right. Left, right.

Dement was so curious, he woke his subject.

"Do you remember what you were dreaming?" he asked.

"Yes," the subject replied. "I was watching a Ping-Pong match."

LaBerge still seems grateful, more than thirty years later, to Dement. Because this seemingly small discovery—that sometimes, in dreaming, our eyes follow the action in the dream—illuminated his path.

The story of how lucid dreaming was confirmed by scientific experiment is well known among lucid dreamers, and LaBerge shared it in one of the Hawaii workshops.

He came into the lab, a doctoral candidate, bent on doing something terribly audacious. He was going to conduct research while asleep. He arrived, changed into comfortable clothing, and Nagel outfitted his head and face in an array of electrodes, including an electrooculogram, or EOG, to measure his eye movements. The rig proved so sensitive that every time LaBerge moved the electrodes spiked—drawing sharp lines on the graph paper ticking in the nearby control room. The bed was uniquely uncomfortable—a gurney, really. And in addition to all these distractions, LaBerge suffered the ultimate performance anxiety.

Yes, he had experienced lucid dreams somewhat regularly. But not on command. So precisely how he might become lucid by wishing it so was unknown to him. He also didn't know whether he would remember the task he had in mind—or if deliberately moving his eyes in the dream would produce corresponding motions in his real eyes. Perhaps Dement had experienced some sort of anomaly. But he lay down, anyway. And on January 13, 1978, Stephen LaBerge had his first signal-verified lucid dream.

Inside his own mind, LaBerge was asleep. He dreamt of literally nothing, standing in a formless void, when suddenly he gained awareness. An instruction booklet for a vacuum cleaner floated by, and as he looked at it, he felt the contours of his own dream body forming. His arms, his hands. There are different levels of lucidity, and sometimes the dream is so vivid, so engaging, a dreamer can forget what task he set for himself. But in the void, with only weightless appliance instructions for company, LaBerge remembered his mission.

Outside his mind, in the waking world, the wires attached to him monitored his brain activity, his muscle activity, and his eye movements. In the next room, Nagel stayed up all night, monitoring the accumulating data. And after eight hours, the dream world communicated to the waking world, in real time. The EOG went into the random, static-like jerks associated with REM. Then, suddenly, it registered two even, controlled movements—the signal LaBerge and Nagel agreed upon.

During the next two years, LaBerge and Nagel repeated the experiment numerous times, perfecting their methods. They realized long, dramatic side-to-side movements of the eyes—in which the dreaming subject looked at his dream ear without moving his dream head—made for the best signals on the EOG. And they brought other subjects into the lab, to prove the positive results weren't dependent on the team of LaBerge and Nagel.

In 1981, LaBerge presented his research findings at the Association for the Psychophysiological Study of Sleep (APSS) and published them in the peer-reviewed journal *Perceptual and Motor Skills*.

This achievement is often portrayed as a triumph. And it was. LaBerge had used modern equipment, unavailable in the days of Saint-Denys and van Eeden, to demonstrate that an experience reported for millennia was actually so.

But the truth is, LaBerge's victory has never been celebrated all that widely.

When he tried publishing his research with more prestigious journals, his papers were rejected. One publication gave no reason for its decision, which is considered poor form; the other delivered the painful truth. "One of our judges said they couldn't find any problem with your study's methodology or controls," he was told, "but they insist *something* must be wrong."

"In other words," LaBerge told us, "they refused to publish my research because they simply couldn't believe it."

Those days aren't entirely over, either. Numerous sleep researchers continue to ignore LaBerge's findings. The authoritative, exhaustive online database PubMed currently includes the 1981 issue of *Perceptual and Motor Skills*. But LaBerge's article is omitted. Curious—was this an oversight?—LaBerge recently looked into the matter. As it turned out, PubMed reviews all the articles it posts, even from refereed sources. "They told me they weren't sure by what criteria they had made the decision not to publish my article," LaBerge told us. "But a decision had been made."

LaBerge rejects any idea that PubMed's choice reflected anything beyond simple bias. "It is a hard science study," he said, "primarily about physiology and not dream content."

He explained, "It is clear, science is a sociological construct, and what gets accepted is in part a function of what has already been accepted."

For the most part, LaBerge seems at peace with all this. But he admits: if mainstream science had embraced his findings, if he received research funding through ordinary scientific channels, he wouldn't be running a workshop in Hawaii.

This might come as something of a surprise.

As it is, LaBerge writes books and serves as the director of the Lucidity Institute. The institute publishes an electronic newsletter, and has inspired a small band of dreamers to aid LaBerge's research. He calls them "Oneironauts" (oh-nigh-ro-knots)—which comes from the Greek and means "explorers of the dream world." He also runs these workshops, one or more per year. At the workshop I attended, he put on a big show, stalking around the front of the room, waving his hands and making pronouncements in mock-dramatic tones: "Ah, but *who* is the dreamer?" he might say, wagging a finger at us, to make us doubt our own existence. Then he would smile his thin smile and raise his bushy eyebrows.

His books have sold well. And the people who follow his work don't merely respect him. They *love* him for what he has done. They feel that when those vacuum cleaner instructions floated by, and he remembered his task—to scientifically verify lucid dreaming—he changed their lives for the better. From a distance, then, it might seem that LaBerge has happily embraced a role as pied piper, leading his followers into the depths of the dream. And he could, easily, be a guru. He could conduct workshops all over the place and pose as the Deepak Chopra of the dream space. But LaBerge just isn't built for the guru business.

As the week wore on, in fact, it became clear how much the workshop wore on LaBerge. He turned up each morning well after his chief assistant started the session. He said his piece. He zipped out. He was nothing if not consistent, turning up late for breakfast, lunch, and dinner, striding into the communal dining area after everyone else had eaten.

LaBerge was fully engaged during the daily workshop sessions. But the effort seemed to take all his energy. Over the course of the week, in between lectures, I got to know LaBerge's chief assistant, a warm, empathic pixie of a woman who goes by one name, Keelin. She has known LaBerge for about twenty years, and she best put the scientist into perspective for me: "Stephen is an introvert," Keelin said. "Public speaking takes a lot out of him. Especially here, where he has to present two lectures every day. He's a research scientist. That's his mentality. But the way things have gone, just staying in the lab isn't an option for him."

In fact, Keelin explained, the advancing years have forced LaBerge to begin considering his legacy. "He's really the only one I know of researching lucid dreaming at his level," she said. "And the thing is, he really believes, as do I, that it can be of tremendous benefit to people. But after him, who will carry it on in the way he has, with such a devotion to scientific methodology? He promotes lucid dreaming because he *has* to."

There was something sad about all this. Like the mongoose, or the dream, it's in LaBerge's nature to run for the shadows. But in his case, he simply cannot have the life and career he wants. And so the experience of Stephen LaBerge provides an important lesson for the rest of us: that sometimes, for all its glorious success, the social construct of science doesn't fulfill on the promise of the scientific method. Sometimes, an entire field of study gets neglected just because it originates on the margins—valuable information that gets stuck in that netherworld between wakefulness and dream.

IIIIIIIIIII

DREAMING ITSELF REMAINS A subject of controversy among researchers. We know sleep is universal. Even the primitive cockroach says good night (mostly during the day), slipping into a kind of stupor for long seconds, its constantly moving antennae going still.

Many fish and reptiles engage in similarly fitful slumbers. But mammals and birds experience regular sleep cycles. And humans? *We* sleep like we were born for it, in ninety-minute phases, descending and reascending through four stages of ever-slowing brain waves before entering REM.

While we dream in some form throughout the night, dreams in REM *do* tend to be the deepest and most vivid. And over the course of the night, our REM phases lengthen—from just a few minutes the first time around, to an hour or more in the last REM stage before we awake. This means, on an average night, we're likely to experience somewhere between four and six REM periods, accompanied by vivid dreams, with the last the most prolonged of all. But this doesn't answer the more fundamental question: *Why* do we dream? Sleep is restorative. But wouldn't it be *more* restful if we weren't seeing ourselves in strange predicaments—taking high school tests naked, or eluding the grasp of roller-skating zombies—in our minds?

For a long time, science has seemed to hesitate in the bedroom doorway, so to speak, wondering if the irrational world of dreaming is a mystery worth solving. No doubt, the link between dreaming and mysticism stigmatized the dream with Paranormal Taint. But the seeming illogic of dreaming, the mystery of its function, is also its allure. Artists take inspiration from the dream. Mystics have long thought the dream a portal to literal other worlds—the Tibetans call it the "dream bardo." For many practicing psychologists, even post-Freud, dreams are a gateway to our subconscious minds, a place where the collision of imagination and memory yields up insight. But for many scientists, the whole search begins along strictly material lines: What evolutionary purpose, they ask, does the dream serve?

There are some potential answers there. Some research focuses on the role dreams might play in memory consolidation. In this view, our sleeping minds identify important information while we get some rest. But in a delicious twist, other research attempts to demonstrate that the dream is important for the exact *opposite* reason. Francis Crick, codiscoverer of the DNA double-helix in 1950, later looked for a second act in dream research. Crick formulated the "reverse learning" theory of dreams, in which the brain essentially goes to the bathroom, flushing away all nonsense and superfluous notions.

If this theory sounds almost personal, reflecting Crick's own opinion that dreams are poop, maybe it was. The subject of dreaming is the ultimate Rorshach test. There are as many interpretations of why we dream as we can dream up. Some think, because we are rested by the early stages of sleep, evolution provided for us to remain at rest as a way to keep us out of trouble—our eyes not prepared for the rigors of staying alive at night.

Jerry Siegel, director of UCLA's Center for Sleep Research, thinks the brain repairs itself during sleep. The dream content is secondary—playtime for the kid while our parent-brains get some real, physiological work done.

Others posit that dreams allow us to practice real-world actions through "simulated threats." In this formulation, dreams comprise a virtual reality machine, where we train for upcoming missions.

The underlying assumption of all these scientific theories is that specific dream content is meaningless. Even the threat simulation theory doesn't hold that we are preparing to be chased by zombies or giant spiders. We just need to rehearse, in case we are chased at all.

LaBerge, however, brings a line of thinking to this ongoing debate that seems particularly useful—in mystical, evolutionary, and practical senses. Working from his own research and that of others in the field, LaBerge subscribes to the theory that *why* we dream, from an evolutionary perspective, is interesting, but perhaps not so all-fired important.

Consider, please, we *did* evolve and we *are* here; so any use we can find for the dream is fair game. In this sense, as individuals we can direct our own evolution and get everything we can out of the human machine.

In Hawaii, workshop participants shared many stories of dreams, lucid and not, that helped them navigate some particularly difficult problem—or simply brought them joy. Gunter, from Germany, said he came to Hawaii to learn how to prolong his lucid dreams. "I get so blown away with the flying," he said, in heavily accented English. "I wake up. I go so fast I cannot even take the time to look at the ground."

Keelin told us that at age eleven, she had a non-lucid dream about her father, who died two years earlier, which illustrated for her that she could continue a relationship with him, if only in her dreams.

I'll share a similar, personal story: in the immediate aftermath of my mother's death, I resisted grieving. I marched through my days like a soldier, always finding some task to perform to keep my mind off what happened. The pain hit when something occurred that I ordinarily would have shared with her. Every time I had that automatic thought to call her, I had to face the realization all over again that she was gone. After a couple of months of this, I had a dream. I was sitting beside my mother, on a wheeled cart rolling slowly down a long, wide hallway. I was sitting up. But she was lying on her back. There was no one pushing the cart, yet still it moved, steadily forward. I looked down at my mother's face. When she died, she was seventy-one years old and frail. In this dream, she looked maybe forty years old and healthy. But she was lifeless. And soon, I realized, I was going to have to get off the cart and allow her to be wheeled down this passage by herself, into death. Immediately, I started to cry. This was a non-lucid dream. But I behaved exactly as I would have chosen. I collapsed my dream body over hers and held on tight. She was still warm to the touch, and suddenly, in the strange manner of dreams, I felt her—there is no other word for it—*loving* me. She was dead. But I felt all the passion and caring she ever held me with as a child, even though her whole body remained still.

Feeling all this love, emanating from her and into me, my dream tears came in a torrent, surging up so strongly, from such a depth, that I wondered if the pain of my grieving might actually kill me. But I submitted to it. And there in the dream, I couldn't really differentiate me from my tears. I merged with my sorrow. And I remained collapsed over her body, holding her, for several more seconds before I woke.

As I half-opened my eyes, all the grief and pain from the dream was there, waiting for me. The hairs on my arms stood up, as if I had just been through a long, passionate embrace. And I cried for a long, long, long time—while, I confess, cleaving tightly to my pillow.

It was one of the worst mornings of my life. And in the ensuing days, I felt spent. But I also realized how numb I had become. And in the wake of this dream, I started to feel. I started to really live again.

I suppose I could have dismissed the dream if I'd wanted to, but I have to ask, *Why on Earth would I have wanted to?*

The dream has long served us, in ways both psychological and scientific. And it's time we acknowledge this.

Remember educational expert Edward de Bono, whom I referred to in chapter 4? He insists that logic is actually a less effective means of problem solving than lateral or creative thinking. This is the source of the cliché—Think outside the box. But this exhortation to escape paradigmatic thinking arises from the truth. And there are numerous examples of times the dream found answers our logical, waking minds couldn't: the chemist Friedrich Kekulé, who discovered the structure of the benzene molecule, dreamt of six snakes dancing in the air together, which finally coalesced into a ring, eating each other's tails in a giant circle. Dmitri Mendeleyev had a dream that helped him establish the periodic table most of us forget from chemistry. Dr. Otto Loewi had a dream that eventually led to his winning the Nobel Prize for his work on the chemical transmission of nerve impulses.

Elias Howe was near bankruptcy, living in poverty, after years of trying to invent an automated, lock-stitch sewing machine. Then he dreamed of being captured by spear-carrying savages, who demanded he create—what else?—a working, automated lock-stitch sewing machine! Now!

The savages of his subconscious carried spears, which Howe noticed had something odd about them; inexplicably, each spear bore a hole, near the tip. When Howe awoke, he designed a new sewing needle. He moved the hole in from the middle of the shaft to a spot near the tip. Good-bye poverty. All hail the dream.

In each case, the dream offered these men solutions *outside* the strictures of thought they had placed on themselves. When they awoke, they ran with these novel ideas, while allowing logic back in as a partner, to create models that worked. From this perspective, people who deny the dream are so in love with their rational minds that they won't let the other guy—the other half of their brains—do its part.

Some readers, I'm sure, might want to reject all these anecdotes. And indeed, many accounts of insight found in the dream are hotly disputed. Psychologist Mark Blagrove argues none of these accounts could possibly be accurate because "the place for problem solving," he writes, "is the waking, social, world."

But fellow psychologist Deirdre Barrett put all this to a test, asking seventy-six college students to spend a week trying to have a dream about some particular problem. Half succeeded in dreaming of the topic they chose, and 70 percent of the successful dreamers reported the dream offered a solution.

I believe the resistance to LaBerge, to lucid dreaming, to Barrett's findings, occurs simply because the dream continues to bear the stigma of that Paranormal Taint, especially when the story involved is like that of archeologist Hermann Hilprecht. In 1893, Hilprecht dreamt an Assyrian priest told him how to interpret two troublesome stone fragments he'd recently found. The priest also told him a third piece would never be discovered. When Hilprecht awoke and looked at the fragments again, he confirmed the dream priest's interpretation was accurate. And no third fragment has ever been found.

Does this or any other dream represent actual communication between the dreamer and someone or something else? Well, it surely doesn't *have* to, so my advice: don't reject the power of the sleeping mind because some might use it to open the door, even a crack, to all the psychic hoo ha that doesn't currently fit into your worldview. Dreams have inspired literary and musical compositions. Some musicians claim to emerge in the morning with entire songs written in the dream space.

LaBerge has spent a long time in consideration of the dream's origin and its uses. And his primary insight starts with a deceptively simple premise: in LaBerge's view, the dreamer is not, as long thought, unconscious. Instead, the dreamer is experiencing the continuation of consciousness—in the absence of sensory input.

He talked about this insight in Hawaii, taking us a bit further down the same path as Andrew Newberg, in chapter 7: human beings do not encounter objective reality directly; we experience the mental model of reality constructed by our minds. In waking life, that mental model is primarily made up of sense-oriented stimuli: we are touching, tasting, seeing, hearing, and smelling machines that only become consciously aware of a little bit of the buffet we're sampling all day long. LaBerge's added insight is that in sleep, the brain *goes on* creating mental models of the world. But with literally nothing to see, touch, or taste, no external input to influence our thoughts, our minds are free to create mental models from our entire internal inventory of memories, associations, and ideas. This is why, in dreams, we encounter everything from chocolate rainbows to flesh-eating zombies.

The thing is, though, the boundary between dreaming and wakefulness is not so thick as we might think. In fact, whichever state we're in, waking or dreaming, the other overlaps.

In waking life, LaBerge explained, the overwhelming amount of input we receive forces the brain to take processing shortcuts in assessing the world around us. As a result, the brain must then fill in the gaps in perception using a mixture of past experience, expectation, and belief about what we're seeing. He illustrated this aspect of the human condition dramatically. He showed us a series of slides in which the brain either eliminates input—say, a black dot—or creates it—a checkerboard square, and we "see" what isn't there.

LaBerge even showed us a piece of writing in which one word was repeated, twice in a row, and asked us to tell him what we read. All thirty-two people in the room *missed* the repeated word the first time, and only after LaBerge asked us to read it a second time, more closely, did even a handful of us see LaBerge's obvious, intentional typo.

In short, our brains prevented us from consciously registering the input we didn't need or expect. "Sooooo," LaBerge asked, eyebrows all screwy, "what is it you're seeing, when you're seeing what isn't so?"

We waited for his answer, like schoolchildren—adult schoolchildren in a class like an LSD trip. "What you're seeing is what you *always* see," he said, "your mental model of the world"—a cocktail, that is, of perception and dream.

That's waking life. But when we sleep, we receive little to no sensory input at all. So what we see and hear is virtually *all* dream stuff—and the inventory of experience at our disposal is as vast as our imaginations. Just as dream intrudes in waking life, however, direct, real-time sensory input can find its way into the dream. That is why we sometimes hear a dream-phone ringing, only to wake a moment later and find our waking phone, really ringing.

Jim, a physicist at the workshop, reported one lucid dream in which his left arm was a rubber baseball bat. When he awoke, he found the same arm pinned under the rest of his body, cutting off his circulation. What happened was clear: his dreaming mind interpreted the sensory input coming from his numb arm, rummaged through Jim's storehouse of mental associations and memories, and emerged, like a child at play, with a rubber baseball bat.

This admixture of reality and dream comprises our entire lives. And LaBerge, as Newberg does with religion, seems to take an operational view of the whole affair. Because dreams *can* be put to good use, why not do exactly that?

LaBerge ultimately asked us to consider using lucid dreams, broadly speaking, in two ways: One, for fun. As Gunter said, in lucid dreams *we can get so blown away with the flying!* During the workshop, the assembled Oneironauts reported lucid dreams in which they flew, walked through walls, and had guilt-free sex with strangers. One woman reported bumping into Jack Nicholson, who wore S&M gear and flashed her a jaunty smile. Another chased down Ellen DeGeneres for free VIP tickets to, quite literally, the *Ellen* show of her dreams.

LaBerge encouraged us to explore these pursuits but typified them as recreational. The other, richer purpose for these dreams is to learn more about ourselves and live better lives.

Keelin, LaBerge's assistant, used her lucid dreaming to prepare for her first appearance as a public speaker. She dreamed of the room she would sit in and the people she knew would be there, then sat down in the dream space and gave her presentation. When she had to do the same thing in waking life, she had already done it once and her anxiety was gone.

Using the same concept of neuroplasticity we discussed in chapter 7, Keelin's experience suggests we can use lucid dreams to rehearse waking life actions—and know that these mental rehearsals will make a real difference. A baseball player or golfer might use lucid dreams to perfect his or her respective swings. A surgeon could use the occasion of a lucid dream to practice a particularly difficult procedure. LaBerge laments that, because lucid dreaming has not been embraced by the larger scientific community, these applications have not been seriously pursued. Still, there is good evidence that mental imaginings bear real-world effects: researchers have found that subjects who imagine performing a physical exercise, including the sweat and strain, enjoyed comparable benefits in strength and physical capacity as the people who *actually* exercised. The increases in strength or dexterity likely arise from rewiring the physical structure of our brain to better accomplish the imagined task.

Neuroplasticity, in a nutshell.

But as LaBerge also teaches, and as I found, just *training* to have a lucid dream is enough to start the process of changing a life, from the inside out.

IIIIIIIIIIII

I HIGHLY RECOMMEND LABERGE'S book, *Exploring the World of Lucid Dreaming*, for detailed methods of how to become lucid. The book is a direct product of LaBerge's scientific research: for the average lucid dreamer, just having a lucid dream or two a week is a worthy goal. But the need to have lucid dreams on command, in order to be studied in a laboratory setting, shaped LaBerge into a kind of lucid dreaming superhero. Avid lucid dreamers speak in reverent tones about his ability to have a lucid dream "any time he wants one."

As LaBerge himself noted to us, repeatedly, his research indicates that lucid dreaming is like any other skill. Some people may have greater natural aptitude than others. But the more the beginner practices, the easier it becomes. I share a couple of his methods here that directly invoke the meditative process—a factor LaBerge himself addressed succinctly. "I understand that for many people meditation might have some religious connotations," he said. "Some of you might be fine with that. Others, not so much. But my advice is, Get over it."

To LaBerge, rejecting meditation or lucid dreaming because of its spiritual connotations is, well, scientifically counterproductive—holding us back from the knowledge and experience that might be ours.

According to LaBerge, lucid dreaming involves attaining a heightened state of awareness in waking life, which naturally bleeds over into dreamtime. And the awareness arises from asking yourself a simple question: *Am I awake, or am I dreaming?*

This question may seem easy to answer, and often it is—but ask yourself,

How do I know when I'm awake? How do I know when I'm dreaming?

Well, for one thing, more *unusual* things happen in dreams. And these events—like flying without the aid of an aircraft—are called dream signs. But the key to lucid dreaming is to become aware of your true state and surroundings as often as possible; and one thing you're bound to discover is how often unusual and interesting things occur *in waking life.*

In other words: Remember the mongoose.

My first morning in Hawaii, that strange brown blur of the mongoose caught my attention, providing a perfect opportunity for me to conduct what LaBerge calls a reality or state test. I must warn you: at first, in practice, the reality test feels awfully silly. But whenever something unusual occurs in waking life, take the time—six, eight, ten, or twelve times a day—to ask yourself that central question: *Am I awake, or am I dreaming?*

Don't answer automatically. Take a moment to think: What evidence would you look for to really answer that question?

Over the years, LaBerge has developed some insightful answers. In a dream, anything you see in print is only written in your mind. So to conduct a state test, read something. Then look away. Now look back at the printed page in your hand. If you're awake, what you see can only be what it was the first time you looked.

But if you're asleep? Whatever you read in the first place was only a figment. And when you look again, your mind can and most likely will retrieve something different from your mental storehouse of written images.

The numbers on a digital watch can turn into words. The words of a great poem might turn into nuclear launch codes.

Machinery also malfunctions quite a bit in dreams, so if you're conducting a state test, reaching for the nearest light switch is a good idea. If you're awake, the light switch is connected to a power source. In a dream, the light switch is connected to anything you might associate with electricity—and, because it's a dream, to absolutely nothing. So if you flip on a light switch and the light flashes like a disco ball, or if an elephant walks into the room, smoking a cigar, you, my friend, are dreaming.

In Hawaii, I cued myself to conduct a state test whenever there was a noticeable shift in the lighting and whenever I saw the brown blur of a mongoose go scrambling by. As I continued talking to a fellow workshop participant or my fiancée on the phone in Philadelphia, I read a nearby sign or the digits on my watch. I looked away, searching the landscape for some detail that might illuminate my condition. Am I dreaming, or am I awake? And more important, I found myself listening to whoever was talking even more intently than normal—really listening to what they had to say and the character of their speech.

The more state tests you conduct while awake, LaBerge told us, the more likely you are to question what state you're in when you're dreaming. And once you ask that question during a dream, it's game on.

The problem is that walking around conducting state tests all day might seem like a big distraction—and at first it is. But after a couple of days, even beginning practitioners will find themselves continuing conversations and silently carrying on state tests at the same time, with no ill effects. In fact, state tests lead to such a high level of awareness that, like meditation, the practice of lucid dreaming quickly leads to a fuller experience of life.

In meditative practice, this is known as the pursuit of "mindfulness"—of being immersed in the experiences of the moment. Andrew Newberg had spoken to his class about this: "When you take a shower," he said, "don't think about your homework or the conversation you just had or are about to have. Think about what you're doing: feel the water on your skin. Think about each action you take as you take it. And nothing else."

The practice also illustrates how alike dreams and waking really are, how many odd or unique experiences are available to us, all the time. I grew to love my state tests, and during my time in Hawaii, some of the most magical, revelatory moments occurred while I was upright and awake.

As a group, we were all alert to the behavior of digital clocks, which always malfunction in dreams. So when the digital clock in our meeting room began blinking, randomly, we all momentarily froze. *Were we asleep?* We all started conducting state tests, then burst out laughing.

I even had a terrific personal encounter with a tree. A cloud passed in front of the sun while I was walking to the general store, the shift in lighting being my cue for a state test. There was nothing around for me to read. And there was no machinery nearby for me to operate—no light switch to flick. So I seized upon the most interesting thing in my field of vision, a small grove of trees. Advancing upon them, I thought, *I will try to magically bend the branches of the nearest tree.* I quickened my pace, still self-conscious enough to be glad no one was in sight, and seized a branch as thick as my calf—maybe fifteen inches around. I didn't believe the thing would bend, but I playacted and heaved to with the raw optimism of the dreamer, unconstrained by the laws of physics. And in a split second, I bent the branch easily—at almost a 90-degree angle. *Whoa!* I thought. *Am I dreaming?*

Just like I had expectations that Steve the Texan Tibetan was a simple hippie, I had expectations of how a tree that big would behave. But in a new environment, those previous experiences had left me with an inaccurate mental model of the world. In my mind, the tree was rigid. But in waking life, the *Ficus Elastica* planted near Kalani's general store bent easily. It was, literally, a rubber tree.

The lesson in all this was profound: lucid dreaming allowed me to engage in a kind of waking meditation *all day long*. It isn't easy. Maintaining that kind of awareness takes some effort. But it's available to me all the time.

One of the workshop participants, Jeff Dalton, captured the benefits best, a few days in. "The mindfulness I get from practicing lucid dreaming," he told the group, "makes me a better father. Things can get out of hand fast with kids because they are so emotional. And I can get caught up in that. But that act of stepping back and reassessing things, of saying, 'All right, what's *really* going on here?' instead of just reacting, has helped me to really *hear* my children, and what they really want."

The whole room responded to this, the energy of the workshop palpably picking up as Jeff spoke. And the truth was, by this point, we were all so engaged with the world around us—well, I think we were *all* high. Tuned in, turned on, and holding absolutely no intention of dropping out.

Understand: this was just the effect of *trying* to have a lucid dream. Actually getting lucid, in this context, is a massive, psychedelic cherry on the cake.

As I mentioned, I had by this time already experienced a lucid dream at home, and it was powerful enough that I knew I was ready to invest the effort it would take to have more.

I triggered that first dream a few weeks before the workshop, using one of LaBerge's other means of incubating a lucid dream. The MILD technique, which stands for mnemonic induction of lucid dreaming, requires the practitioner to focus or meditate on a specific thought: *The next time I'm dreaming, I'll remember to recognize that I'm dreaming.* One effective way of practicing this technique is by reciting this mantra and also remembering a dream in as much detail as possible. I used a recurring nightmare, imagining myself really *in it*—recalling all the images and sounds and associated emotions. Then, using LaBerge's instructions, I'd imagine becoming *aware* that I was dreaming during some part of the dream—and what I would do as a result. This, too, is a meditative practice. And in my case, the nightmare I chose to meditate on had first appeared about twelve years earlier.

In this dream, I was home alone when a man's face appeared in the window. He saw me notice him and receded slowly into the dark, his face disappearing like a fish tickling the surface of the water and retreating into the depths. But a moment later, his face emerged from the dark again. Each time I had the dream I was terrified by this point, wondering, *Who in the hell is this creepy dude?*

The way the dream usually went, he eventually moved away from the window and knocked on my front door. Slowly. Each beat of his hand on the wood was interrupted by long seconds of silence. The dream would always end the same way: after a few minutes of his knocking, I just got pissed. *Who is this guy?* I'd think. *Trying to terrify me like this?*

"Come on!" I'd start yelling in my dream. "Come get some of this. You're gonna kill me? *I'm* going to kill *you*. I'm begging you to come in here, so I can murder you. I *dare* you to come in here!"

I'd had variations on this dream maybe two dozen times or more, and every time, before long, I was double-dog-daring this boogeyman to come into my dream house so I could beat him to death. And each time, he obliged me. Suddenly, he kicked the door open. We flew at each other in a rage. And just as we began to grapple, I woke up throwing punches, my right hand balled into a fist.

Using the MILD technique, I imagined all this repeatedly, only instead of becoming filled with fear or anger at the sight of the man's face in my window, I imagined myself realizing, *This is a dream.*

From just this part of the exercise, I quickly realized something startling: I'd had lucid dreams in the past, during nightmares. But in those dreams, when I realized I was dreaming, I used that incredible power . . . to wake myself up. LaBerge consoled me, saying this is a common reaction. Just not terribly helpful. "When we realize it's a dream," said LaBerge, "the appropriate response is to take that seriously. The boogeyman chasing you has no external reality. You don't have to wake up or fly away. In fact, the best results people get are when they react to whatever's chasing them with curiosity and compassion."

And so in the next stage of MILD, I started imagining more fruitful responses than abject fear and murderous rage. And after a few days of meditating on this, maybe twenty minutes per day, I saw the bastard. I was asleep and there he was, at my window, leering. *I've had this dream before,* my dream self thought. And that was all it took to bring me the awareness I needed. For the first time ever, I had real lucidity in a dream.

The walls of my apartment seemed to grow more stable and more vivid. The floor underneath my feet felt solid. *Dear God*, I thought. *This is just like* The Matrix. I felt as if *I* had just appeared, in some alternate reality. I could even feel the pads of my fingertips tickling my palms as I nervously considered balling my hands into fists.

Feeling in control of my fear, I went to my front door—and opened it.

A few seconds later, my tormentor appeared. And to my surprise, he was not particularly malevolent looking. In fact, he looked like any ole guy at all—a generic beer-drinking dude. But he was having none of this new Steve. He filled the doorframe, looking at me quizzically, as if to say, *Hey, this isn't what we do . . .*

But I held my ground. I said nothing.

His face took on a sour cast then, like he was offended that I wouldn't play along. He reached into his pants pocket and pulled out a gleaming black handgun. He paused for a moment, the gun pointed toward the floor, and regarded me carefully—to see if my reaction changed.

To say that the moment *felt real* describes nothing and everything. In normal dreams, images retain a kind of washed-out quality. In lucid dreams, conscious awareness can bring out every detail. LaBerge says this is because we're not just seeing objects, as we do in waking life, but every image we've ever seen of the object, every idea we've ever had about the object—and all our emotional associations, besides.

I could smell my own sweat, and his, my apartment suddenly musky as a locker room. I could see where the serial number was etched on my mental model of his gun.

I was lucid, but didn't even think to ask the boogeyman any questions. Like *Why are you here?* Or *What do you want?* The kinds of questions LaBerge recommends. The truth is, I was too frightened to think of anything to say. But I held on to the thought that, because this was a dream, no harm would come to me.

He lifted the gun from his side. He pointed it at me. I could almost feel the barrel, somehow, from six feet away, aimed directly at my chest. He paused. He looked right into my eyes. And I had the sense he was giving me one last opportunity to panic, to slip back into my usual reaction.

I felt a sudden upswelling of fear, like bile in my mouth, but steeled myself. *This,* I said, *is a dream.*

He'd had enough. He pulled the trigger, the sound enormous. Each explosion filled and colored the room where we stood. But when I looked down at my chest, nothing much was happening. My shirt moved a little, as if blown by the wind. I felt no pain, saw no blood. And I experienced . . . elation, a thrill that lucid dreaming really works, and that I had just now stood up to my own fear. I had just stood up to bullets.

After the sound of the last gunshot faded, he stood there for a moment, still pointing his smoking gun at me—the only sound now the click of an empty chamber. Then he let his hand drop, suddenly, to his side.

We stood there for a second, in silence. And then he did something I never would have expected. He grinned at me. I let my own tension out with a laugh, and he started laughing, too. Suddenly, we were buds. *Amigos*. My personal dream terrorist, my friend. He nodded and a wordless communication passed between us: *Congratulations,* he seemed to say. *You got it.*

I woke up, feeling like Superman. Feeling like Neo. And spooky dude? He hasn't been back. No more mornings waking up tense. No more mornings colored, even momentarily, by the natural anxiety associated with a recurring nightmare.

What Hawaii still had in store for me, however, proved even more powerful than that. What I learned in Stephen LaBerge's lucid dreaming workshop shed a new light on the paranormal—and even set to rest the Family Ghost. But before we come to the close of this tale, there is one more story I need to share.

9

After-Death Communication?

How a New Therapy Uses the Dead to Help Patients Live

When I look at the religious question as it really puts itself to concrete men, and when I think of all the possibilities which both practically and theoretically it involves, then this command that we shall put a stopper on our heart, instincts, and courage, and wait—acting of course meanwhile more or less as if religion were not true, till doomsday, or till such time as our intellect and senses working together may have raked in evidence enough,—this command, I say, seems to me the queerest idol ever manufactured in the philosophic cave.

—William James, *The Will to Believe*

Tom Lareau broke on a hillside in Dak Do.

A Claymor mine exploded, wounding his best friend. Lareau reached for him and saw that his friend's legs were almost entirely blown off. Only some tendons and a thighbone bare as rock still descended from his hips.

Lareau picked his friend up, slung him over his shoulder, and started up the hillside, marching toward the area where Evac helicopters had already begun to descend. On the way there, the backs of Lareau's legs grew increasingly wet and warm. And the wounded man hollered in his ear, "Call my family," he said. "You tell them what happened. Take care of my family. Don't let me die."

"By the time I reached the top of that hill," Lareau told me, roughly thirty years later, "I had lost my mind."

By the time Lareau reached the top of that hill, his friend was dead.

Lareau, like so many Vietnam veterans, carried the traumatic events he witnessed in war inside himself for decades. He worked to the exclusion of all else. He wanted no friends, and the world obliged him. He never contacted his friend's family, because the prospect of facing that hillside again was just too tough. He even tried to kill himself on two occasions. Finally, in 1995, he wound up at a Chicago area Veteran's Administration medical center. There, he met a doctor by the name of Allan Botkin.

Botkin provided him with an experience Lareau still can't explain. All he knows is—he is finally content. "I saw my friend. He was sitting on a rock in a clearing in the jungle. He told me he was okay. He was in a beautiful, peaceful place, and he told me it was all right that I never contacted his family. He understood."

Lareau does not know if he received a communication from his friend or merely enjoyed a vivid, pleasing vision—a fantasy that healed him. Botkin doesn't claim to understand the experience all that much better. But he does claim he can induce similar experiences in others. In fact, Botkin claims that he and the therapists he has trained have induced the experience in thousands of people already.

I heard Botkin promoting a book about his procedure on the radio and called him shortly thereafter. At this point, my mother was alive, her cancer not diagnosed. But my brother had died more than a decade earlier, and my brother-in-law had passed away just months before. I decided to tell this story here, out of chronological order, because Botkin's odd therapy draws upon virtually all of the issues raised in this book—issues associated with consciousness, belief, dreamlike states, the relationship between mind and body, recovery from grief, and our current understanding, or *mis*understanding of the paranormal. And frankly, I think I understand it better now, after all my subsequent research, than I did at the time.

Botkin had told me, over the phone, that he wasn't claiming to put people directly in contact with their deceased loved ones. But he wasn't claiming *not* to, either. "I don't know what's happening, exactly," he said. "But I do know it's safe and it's effective in helping people deal with their grief."

Botkin and the retired soldiers I ultimately met with were on something of a mission. Thousands of American soldiers, they said, returning from Iraq and Afghanistan, face more danger at home. Post-traumatic stress disorder (PTSD), grief, guilt and loss, addiction and recovery—all lie in wait, roadside bombs threatening the rest of their lives. "I know Dr. Al can help them," one of Botkin's soldiers said.

Five years later, I think that Botkin's therapy might be more complicated than either believers or skeptics would at first admit, and that some concerted effort should be made to understand his strange procedure. But I also know: the only way Botkin will get a wider hearing, is if we reframe our view of the paranormal.

IIIIIIIIIII

IN PERSON, AL BOTKIN is a big man, well over six feet tall, his hair close cropped and blond, his shoulders broad, his manner matter of fact. A former athlete, Botkin pursued a career in basketball before he blew out a knee. And when I met him, he still carried himself with an athlete's physical confidence and upright posture. "I'm still surprised that all this happened," he told me. "I know, everybody says this. But I am the last person I ever thought would start a therapy like this one. I've never been, and still am not, a 'New Age' guy. But this started happening, and it seemed to be helping people. So I went with it."

Botkin calls his therapy "induced after-death communication," or IADC, a moniker he admits is paranormally loaded. He defends his naming decision, pointing out that it is an attention getter and it accurately depicts the way most patients describe it. Personally, I think he may have doomed himself with this name; he took a therapy that might already bear some Paranormal Taint and further lathered it in hoo ha. Certainly, the fact that his therapy has grown in the five years since I met him but has never come close to catching on might be taken as evidence of this.

Whatever the name, certainly, Botkin was always going to face an uphill climb. His therapy originates in an unusual treatment, called "eye movement desensitization and reprocessing," or EMDR. Like Botkin's IADC therapy, the discovery of EMDR has a kind of mythic quality, tied up in a single unlikely event. In 1987, Dr. Francine Shapiro, a trained psychotherapist, had physically recovered from a bout with cancer. But as she walked around a small lake, on a spring day, she ruminated on her own anxiety. The sun shone, but her disposition was gray. And then, suddenly, she simply didn't feel the weight of anxiety anymore. Good feeling had returned to her with a haste she found jarring.

This might sound like a kind of non-happening to most people; certainly, it did at first to me. Mental states pass. That is their very nature. But Shapiro was deeply concerned with this subject—concerned for her life.

In her book, *EMDR*, Shapiro describes how her doctor had warned that her cancer could come back. How to prevent it? He had no answer. When she discovered that both her own cancer and the colitis that killed her sister had potentially been linked to stress, she resolved to become her own doctor; and so, by the time she found herself near this lakeside, she had spent a lot of time exploring the relationship between mind and body, in hope of helping other people and herself.

She worked as both researcher and subject, continually monitoring her own body for links between mental states and physical effects. And so, having suddenly felt her depression lift, with no explanation she could find, she dwelled on the sudden change as she walked by the lake. What had even been on her mind, previously, that troubled her? Sure enough, she found the source of her negativity again, but incredibly, she discovered, those thoughts had lost their emotional charge.

She was even more intrigued now, and she had the self-awareness necessary to notice that, as she recalled these bleak ideas, she felt her own eyes move spontaneously back and forth. She began experimenting. She riffled through a mental inventory of emotionally charged memories and depressing thoughts. And as she did so, she purposefully moved her eyes side to side.

Over and over, she noticed the same effect. The depressing thoughts took hold, then quickly lost their power.

She spent the next many weeks asking friends and acquaintances to go through the same procedure. And as she heard their reports, she developed a protocol to elicit the best results.

The method she came up with, dubbed EMDR, was both simple and strange seeming—combining both psychology and physiology. In phase one, the therapist and client talk about some traumatic memory and the client is encouraged to reflect on the images, sounds, or emotions associated with it. Then, the physiological component of EMDR is initiated. In this next phase, the therapist asks the patient to follow the movement of some object—strictly with his or her eyes. "Don't move your head at all," Botkin told me, when it was my turn.

The idea is to get the patient's eyes to move, left-right, left-right, as if they are witnessing a Ping-Pong match.

Individual therapists use different objects. Some raise the index and middle fingers of their hands, moving them side to side directly in front of the patient's face. Others use an electronic gizmo that sends a single pinpoint of light pinging back and forth.

Shapiro began publishing on EMDR in 1989, and a new therapy was born. Or was it? According to Shapiro herself, in an updated edition of her book, though there are some theories under research, the exact mechanism by which EMDR works remains something of a mystery. So I digress, for just a moment, to ask the question: Does this make EMDR paranormal? I ask because EMDR faced a similar level of stigmatization.

Therapists had been working with a variety of treatments for many years, in the wake of natural disasters and wars, to help individual patients get their lives back. But the effectiveness of varying post-traumatic stress therapies has been notoriously difficult to sort out, and some PTSD cases have simply seemed intractable. So, as might be expected, many in the psychological community were deeply suspicious of Shapiro's odd origin story—the lady by the lake, suddenly glimpsing this steady, passable bridge between mind and body. But as Shapiro herself puts it, the main criticism seemed to be that EMDR was "too good to be true."

The debate raged among U.S. psychologists throughout the 1990s; the Wikipedia page on the subject provides a quick view into the assertions and counter-assertions that continue to this day. But for now, it seems, EMDR is winning: numerous papers demonstrate that the therapy helps patients and has a measurable impact on brain function. And the practice of EMDR is gaining adherents, not losing them. Shapiro's EMDR Institute estimates that 100,000 practitioners worldwide have been trained in EMDR, which has been endorsed as an effective treatment for PTSD by the American Psychiatric Association, the Department of Veterans Affairs/Department of Defense, and the International Society for Traumatic Stress Studies.

Botkin himself even refers to the time "before EMDR" as "the bad old days." He spent the 1980s, before Shapiro had published her work, working at the Veteran's Administration (VA) Hospital in Chicago, where exposure therapy was still the standard treatment. "We had these guys, traumatized war veterans, watching *war movies,*" he said, laughing darkly. "I mean, we sat them down in a room and turned these films on with scenes of the most awful violence. And we told them, when they wanted to leave, to *stay* because it would make them better."

At that point, in his own estimation, Botkin's career amounted to ten years of failure. "It was rare, in my experience, to see a patient really get better," he said. "The VA was a depressing place. And we told ourselves that it wasn't us, or our treatment. It was our clients. They were just so damaged in the war. . . ."

But for a sense of professional desperation, Botkin was an unlikely practitioner of EMDR. He trained as a behavioral therapist, earning his doctorate in psychology in 1983 from Baylor University, and for a long time he believed in the central tenet of behaviorism: its singular focus on the way we respond to stimuli, its insect creed. "People's inner lives," Botkin told me, "their thoughts, seemed entirely irrelevant to me. I thought you just had to look at their behavior."

His grip on that worldview was weakened, however, by years and experience. Dozens and then hundreds of men paraded past him, retired soldiers awash in alcoholism, bar fights, domestic violence, divorce, drug abuse, total estrangement from their families and society—and all because of what they reported happening *inside their heads*. Flashbacks, depression, despair, guilt, homicidal or suicidal rage. "In the bad old days, before EMDR, once you got a soldier to really open up to you about a trauma, he didn't sleep for days," Botkin told me. "They bounced off the walls."

Then, one day, a colleague came in with a paper written by Francine Shapiro, extolling the virtues of EMDR and its odd protocol.

Botkin and another therapist listened intently to the psychologist describe the basic outline of Shapiro's therapy.

And then they laughed.

The giggle factor can arise, it seems, whenever we hear something that doesn't fit with our worldview. And Botkin admits, when he first heard about EMDR, he laughed it up pretty good. "In that paper," Botkin said, "she describes some fairly fantastic results. And we laughed like crazy. The idea that something so simplistic—you know, waving your finger in front of somebody's face—would have that kind of effect? We *joked* about it."

In fact, between guffaws, Botkin barked out laugh lines like, "Which finger do you use?" before exploding again into mirth.

For many, many thousands of therapists, and hence for many, many millions of clients, the matter ended there. With a doctor feeling superior enough to laugh in the face of new, strange information. But Botkin and his colleagues decided to look further into the subject. "What did we have to lose?" Botkin remembers thinking. "We figured it was worth a shot."

The year was 1990, and Botkin and his colleagues came back from a seminar, filled with uncertainty about whether this new method would work. They had been dealing with the same clients, in many instances, for years, with no relief. But EMDR turned the whole place around. "These guys used to be gripping the arm rests on their chairs at the end," said Botkin, "white knuckling, and we'd have to say, 'Time's up.' Now they were going through these painful memories, and at the end they were relaxed."

Botkin used to worry about what would happen to the men in his care after they left the office. Now they left him saying things like, "Thanks, doc. I'm gonna go take a nap."

Those first weeks at the Chicago VA, post-EMDR, were heady stuff. Botkin and his colleagues raced down the halls after nearly every patient session, going into each other's offices, closing the door and engaging in joyous high-five sessions: "After all these years, where I hate to say it, we might have done more harm than good," Botkin told me, "to see these guys getting better was astounding. It just felt great, to finally *help* these people."

His own therapy was an outgrowth of all that. And, like Shapiro's development of EMDR, it happened because he attended to a kind of accident.

||||||||||||

BOTKIN WAS ALONE IN his office with Sam, a vet in the throes of depression over a trauma that had occurred twenty-eight years earlier. By now, Botkin had been using EMDR as his main therapeutic technique for maybe five years. He had not fully resolved his every client's every source of trauma. Where war vets are concerned, traumatic memories can be so numerous that it takes considerable time to find and deal with each one. But Botkin had used EMDR to help Sam confront and resolve painful memories; and, finally, what seemed to be his core trauma broke to the surface.

The old solider was sitting there, blubbering in Botkin's office chair, sharing with him the kind of tale that seemed to come straight from a horrifying war flick. Sam had, during the war, befriended a ten-year-old orphaned Vietnamese girl. He planned to officially adopt her and bring her home to the states. Then word came: all the orphaned children on Sam's base were to be transported to a Catholic orphanage in a distant village.

Sam felt devastated at the prospect of being separated from the girl but dutifully loaded her onto a flatbed truck in preparation for transport.

Then, suddenly, shots rang out.

A sniper or snipers were firing on the base.

Sam and the other soldiers present started grabbing kids off the truck and pushing them down on the ground, shielding them with their bodies. Minutes later, after the firing stopped and the crisis was over, Sam looked around for the girl.

He didn't see her till he walked to the back of the truck and found her—face down on the ground.

The truth dawned on him by degrees.

She was motionless.

There was a small spot of blood on her back.

He grabbed her by the shoulders and turned her over.

And then he saw: the whole front of her abdomen had been blown away, torn apart by the bullet that entered her from behind.

Sam clutched at her lifeless body, and eventually his fellow soldiers had to separate him from the girl.

Botkin pulled this whole story out of Sam, then asked him to stay with the feeling of grief provoked by the telling.

The first phase of EMDR was complete. Now it was time for the physiological step. Like doctors dispensing medicine, EMDR practitioners speak of "administering" an eye movement. What this means, in practical terms, is that they set some object to moving, side to side, a few feet from the patient's face. By this time, Botkin had chosen as his instrument of choice a long, white stick his clients dubbed "Dr. Al's Magic Wand."

Botkin waved that wand in front of Sam, watching the old soldier's eyes track the tip dutifully, side to side.

At first, as expected, Sam's sadness *increased*.

"Stay with that feeling," Botkin instructed, then waved his wand again.

They continued on this way for a while, and Sam's sadness slowly began to decrease, like a deflating balloon.

Toward the end of their hour-long session, Sam's face was wet with tears. But he expressed relief. He had gotten through his story. And he felt the relaxed after-effect associated with EMDR.

Normally, Botkin would have broken off the session at this point—their work done. But this time, spontaneously, he administered one more eye movement, with no specific instruction. He says now that he thought of the extra eye movement as a "kind of treat. You know, 'this has been good for you, so here's one more.' Like dessert."

He just waved his wand in front of the soldier's face, with no further instruction. Then he told him to close his eyes.

Botkin figured this was it, the end of the session. But then Sam did something surprising: eyes shut, he smiled. Broadly.

Botkin sat there, watching Sam intently. And then the soldier did something that disturbed Botkin: he giggled.

Botkin was thrown.

EMDR had the effect of relaxing patients, of helping to settle them down and stabilize their emotions. But this was a massive mood swing.

He waited expectantly, then Sam opened his eyes again and told him, he *saw her*—he saw the orphan girl. Sam was smiling now, clearly euphoric, and the story he told stunned Botkin.

Sam claimed to have seen the girl as a beautiful woman. She thanked him for taking care of her. She seemed happier and more content than anyone he had ever seen.

"I love you," he told her.

In response, she embraced him and said she loved him, too. "I could actually feel her arms around me," Sam claimed.

By now, Botkin was scared. "I assumed that the agony of his grief had somehow produced a hallucination based on fantasy or wishful thinking," he writes in his book, *Induced After-Death Communication*. "I had never witnessed or heard of such a response during psychotherapy . . . If Sam had hallucinated, the intense stress of his traumatic memories had somehow compromised his ability to differentiate reality from fantasy. That worried me."

Over the next three weeks, however, five more of Botkin's patients spontaneously claimed similar experiences, "all with the same reported vividness," writes Botkin, "the certainty the vets expressed that it was real, the positive assurances they reported from the person who died, and the unprecedented resolution of long-standing, intractable, traumatic grief."

Botkin had come to expect that patients would leave his office after EMDR with less sadness. But after these unusual sessions, his clients left his office feeling joyous.

Botkin pored through his notes, looking for some common thread running among these patients. And after some time, he found one: each of these six patients had received an "extra" eye movement. There were other changes he had made over the years in the EMDR protocol, all now incorporated into his IADC therapy, but this "extra" eye movement seemed particularly important.

"Usually, you're telling them what to focus on," Botkin later told me. "But in these instances, in my notes, I could see I had given them a final eye movement with no direction."

This first "treat" he had dispensed to Sam had been inadvertently replicated. And all of these patients thought Dr. Al's Magic Wand had granted them temporary passage to communicate from this life—across the boundary of death. Botkin had been trained not to dissuade his clients of their personal beliefs, unless the belief seemed to lead to some direct harm.

As Botkin put it to me, if a patient thought he could go up to the VA roof, leap off, and fly, he would have counseled him otherwise and restrained him if necessary. But this was different. He continued to monitor the patients who spontaneously had the experience. They all seemed to be doing well. None seemed disassociated in any way from reality. In fact, they seemed to be *reconnecting*. "I don't consider anything that goes on in IADC to comprise scientific evidence for the afterlife," Botkin told me. "But once I saw that it was safe, I wanted to explore it as a therapeutic tool."

He had seen great improvement in his patients with EMDR, "but this was on an entirely different level," he says. "People were leaving my office not just more grounded and peaceful, but happy. Even ecstatic."

So he took the next step and started giving his clients one last eye movement, without direction, as part of his own protocol. And incredibly, they had the same experience. They met deceased friends, loved ones, even enemy soldiers they shot. They felt them, smelled them, and believed the experience, in almost every case, to be real.

Botkin told me all this in my hotel room. And I found myself unable to believe a word of it.

Then he opened his briefcase.

It was my turn.

IIIIIIIIIIII

I HAD BY THIS time talked to several of Botkin's soldiers on the phone. And I knew what to expect. But when Botkin reached into his brief case and pulled out . . . a magic wand? I laughed. The giggle factor kicked in, big time.

Botkin smiled, waiting patiently for me to compose myself. This was, in fact, serious stuff. My brother-in-law had died after a protracted battle with cancer just a few months earlier. I had sat in his hospital room for many long nights, including the night he passed away. And I'd arrived in Chicago not just as a journalist, but as a client.

I often woke in the middle of the night, with the smell of that hospital room heavy on my face, like a rag clamped over my nose and mouth. The sense of helplessness I felt watching him go would come storming back into my consciousness. I would sputter, get up, and walk around my apartment—unable to breathe normally until I gave into my grief for a while. And the same sensation came at me at other times, too, usually after I had spent the day out in the busy, heavily populated streets of Philadelphia. Suddenly, alone, unlocking my apartment door, in the evening, I would suddenly catch that same smell, boiling up out of my own subconscious. Then the rush of emotion came—the same cycle of helplessness, panic, and despair. The only thing I'd found to do to help myself at that stage was to walk the sensation off, like a football player who just tweaked his knee. But these episodes were wearing me down. There were too many nights in which I simply lost sleep. And I had, in fact, experienced one of these episodes the day before my flight to Chicago. So IADC, EMDR—the acronyms mattered to me only as a journalist. As a person, I was hoping Botkin might help end these flashbacks.

In my hotel room, Botkin asked me to focus on the smell and the images and the sounds that were the source of my discontent. He listened, wand in hand, till I felt good and weak from the weight of these recollections. He worked with me verbally, like any cognitive therapist, trying to reach the depths of my sadness. Only after I leaned heavily forward in my chair, did he seem satisfied. "Stay with the feeling," he said, then lifted his stick in the air and moved it across my field of view, repeatedly, for maybe ten seconds.

The act of doing first-person journalism sometimes feels incredibly silly. And at the time, this episode with Botkin ranked right up there. But I pressed through any self-consciousness, and in the first stage of the process, nothing terribly dramatic happened—or at least, nothing shot through with mysticism. Botkin administered maybe six eye movements, and by then, incrementally, my sense of sadness and despair had lifted. I felt relaxed and even yawned. I started thinking about Botkin's departure and seriously considered taking a nap.

"Great," said Botkin. "Now I'm going to give you one more eye movement—without direction."

A note here: in the beginning, when Botkin's soldiers first started reporting their experiences, the whole thing happened organically. But I had of course been given a direction before I ever arrived. I had listened to a radio interview Botkin conducted. I had spoken to him on the phone. And I had just interviewed him in person. I knew what *he* expected of me now was that I would have a vision of my brother-in-law. That is an awfully powerful suggestion, so in this respect I knew I wasn't an ideal research subject. But for whatever it's worth, over the course of two days, Botkin performed the procedure with me twice—and each time, well, something happened.

The first time the vision I had was light, relaxed, like an incredibly vivid daydream. My brother-in-law appeared to me as he had in his early thirties, with long black hair and a full, healthy face. He appeared close to me, in the pitch dark.

"You're all right," I said, in my mind.

"I'm good," he told me, and laughed.

This man had been in my family since I was twelve years old. He was like a brother to me.

"I love you," I told him.

"I love you, too," he said.

Then suddenly, he appeared before me at a distance, swinging a Wiffle Ball bat, like he did when he played with me as a child. "All the times we had," he said, "don't end. They go on forever. They're still happening."

I wouldn't have expected my brother-in-law to reflect with me on the nature of time. But I opened my eyes after he said "I love you" again. That first vision was over. The most logical explanation was that Botkin's ministrations rendered me receptive to the more positive memories in my own subconscious.

I say this because, even though no one has yet pinned down the exact mechanism for EMDR's effectiveness, the most promising area of research relates to the storage of memories. The side-to-side eye movements of EMDR recall the more chaotic motions that occur in the rapid eye movement stage of sleep. And some sleep researchers believe one function of our vivid, REM-stage dreaming is to help us consolidate, process, and file our memories—an internal historian, hard at work, contextualizing and writing down the relevant aspects of our lives.

The intriguing twist here is that, studies demonstrate that the quality and quantity of REM sleep declines dramatically in people suffering from PTSD. Is this because some events are so traumatic we literally can't bring ourselves to process them? Unprocessed memories might explain why soldiers suffering from PTSD react to the memory of gunfire with a full-out fight-or-flight response—gripping their chairs or diving to the floor. This might also explain why my subconscious occasionally kicked the smell of my brother-in-law's hospital room *up*, into my conscious mind; I had yet to make the experience a part of my past, and so there it was, in the present.

If all this is true, the waking, directed eye movements of EMDR facilitate a physiological process in the brain, similar to REM sleep, and get the process of memory storage moving again, putting traumatic memories where they belong—in the past.

"You won't know what kind of effect this had until later," Botkin advised me.

But I already knew one thing: seeing my brother-in-law swing a Wiffle Ball bat, and run after he hit the ball, felt invigorating after last seeing him unconscious and in bed. I felt lighter, like I had a new memory of my brother-in-law now, sitting between me and all those darker visions. Looking back, I wonder if processing the negative memories on which I was stuck cleared a path for more positive associations to come back to the fore.

The next day, Botkin put me through the same steps. We talked about my brother-in-law. But when I closed my eyes, I saw my oldest brother. He had died more than ten years earlier. But I'd been thinking of him that morning, so perhaps the power of suggestion had already been at work.

He suffered horrible acne as an adolescent, leaving his face as an adult still pitted and scarred. He had contracted diabetes many years before his death, and he lost a lot of weight he never regained. But in my mind's eye he appeared whole and strong and healthy again—his weight up, his face unblemished. I was surprised to see him this way, so different but still looking just like my brother. And this episode, like the one with my brother-in-law, seemed marked by that interplay between my conscious thoughts and sudden, spontaneous happenings.

We exchanged "I love yous."

We embraced.

I asked my brother if he liked it where he was, and he said, "It's great."

Then just as I'd seen my brother-in-law suddenly appear with a bat in his hands, I saw my brother standing there with a guitar. He used to play electric guitar when I was a child, and I can still remember arriving home from a trip to the grocery store with my parents to the sound of then-current, now-classic rock, exploding out of his bedroom window. These Herculean, arena-rock riffs, played without drums, vocals, or any great talent, marked the soundtrack of my childhood. But in this vision, when he played a chord, an array of colors flew from the strings. I can't say I recognized the song. But the sensation was exhilarating.

I wasn't certain what brain process triggered this. But I liked it. Still, my skepticism created a gulf between me and Botkin's soldiers.

We sat together for somewhere between three and four hours—me, Botkin, and six veterans of the Vietnam War. Botkin had secured us a big conference room for the day, and I sat and listened as each soldier took turns sharing his individual story. I wondered if the people *not* speaking might be bored, being forced to sit through everyone else's tale. But the camaraderie they shared was clear. They liked being there in the same room with each other, *for* each other; and a couple of times, when one of them started to cry, the others urged him on. "Don't bury it," they might say, or—echoing Botkin—"Stick with that."

The amount of misery these men suffered, not just in war but for decades after arriving home, was staggering. As hard as they tried, they simply could not turn on all the old feelings and behaviors that comprised their former, civilian identities. They hurt, and they self-medicated. They drank, smoked pot, or turned to harder drugs like coke and heroin. They tried to talk to their mothers, their fathers, their wives or kids, but they felt as if they were hollering across a chasm and into the wind, their voices lost in the space between. But they at least understood each other—their fellow soldiers. And so they came to the VA, and the best relief they got, for many years, was just being around other similarly wounded men.

"I see another vet," one soldier said, "and I don't even have to like him, as a person, and I *love* him. You know what I mean?"

Another told me, "I'm not saying this to scare you. But if I had to kill you right now I could just turn off all my feelings and do it, like a job, and not feel anything."

I had heard similar statements before, from ex-cons, who were trying to tell me what prison life had done to their minds. So I was neither shocked nor scared when this soldier talked about how easy it would be to kill me. But the next part moved me. "I don't like that," he said.

To a man, they said the closest they've come to normal was through "Dr. Al's Magic Wand."

EMDR provided them the first real progress they had felt since coming home. But IADC took them to a whole different level of healing. One solider said he used his IADC session to resolve his differences with his mother, who died feeling estranged from him after he returned from the war; he still felt pain at the loss of her, but IADC had allowed him to acknowledge that pain. Another vet told me he spoke to an enemy solider he had killed, and, incredibly, the enemy forgave him. "He's in a better place," he said.

In the years since, these men learned how to hold down jobs and feel optimistic again. They still cried sometimes over the past. But now they felt in control of the experience rather than subsumed by it.

I waited a long time before I popped the paranormal question. But eventually, I had to ask, "So, do you feel these visions were real, that you were really visiting with the spirits of the people you saw?"

"I *know* it," said one soldier, immediately.

Another said, "I can't prove it, but I believe it's real."

Then a big bear of a soldier interrupted: "Dr. Al said he gave you the stick. What do *you* think? Was it *real*?"

I understood that the most parsimonious and logical explanation for my visions was that they were some form of spontaneous, waking dream—an imagining triggered by the power of suggestion. I also knew those visions *felt* great and seemed to provide me with fresh, positive memories of two people I had lost. Today, I wonder if those visions are so clear in my mind because EMDR does mimic the REM stage's role in memory processing. But none of that mattered to me then. What mattered to me was that I had just listened to this group of soldiers tell me their painful stories for several hours, and I was not about to come between them and a hard-won sense of peace.

In this book, we have already learned that the brain mediates all experience. Divining the objective reality of what the brain shows us is harder than we think. Call that my intellectual out, if you like; or simply call me a coward, if you're of a mind, for failing to raise my own sense of disbelief. But more than five years later, I remain comfortable with how I responded. Strict philosophical materialists usually hold that all paranormal belief is harmful. But that isn't what I found in talking to the men at that table. So I had one job that I could see: to get the hell out of their way, as fast as possible.

"I don't know," I told them. "It was such a strange experience. And I'm still trying to process it."

"Give it time," one of the soldiers advised me. "You'll see."

I nodded and moved on. And in time, I did see. I never did have another flashback to my brother-in-law's hospital room. And even to this day, when I think of my brother, or my brother-in-law, I no longer first see images of them in failing health. I see the men who appeared in the wake of Botkin's wand. I see the men from my IADCs.

This is not an argument for the objective reality of what I saw. This is not an argument for life after death. But it is an argument that we might do better to set aside some of these epistemologically unanswerable questions from time to time. It is an argument that, at least in the case of Botkin's soldiers, sometimes paranormal belief simply *works*. Whether it points to an objective reality or not.

There have been no well-researched, controlled studies of the effectiveness of Botkin's method. He has been trying to get that kind of project together for many years, looking for some independent researcher to come in and assess whatever it is he's wrought. He has told me, at various times, about psychologists and institutions that have expressed an interest. But as of the summer of 2010, he is still looking.

Is IADC effective? In a scientific sense, we don't know. In an anecdotal sense, hell yes.

Botkin, in the absence of independent research, has now trained sixty therapists in eight countries who are using the IADC procedure to help patients deal with trauma and grief. I spoke to a few of these therapists, who felt IADC was working for their clients. And Botkin's soldiers have remained willing to promote IADC, largely because they want soldiers returning from Iraq and Afghanistan to hear about it. "I went many years till I got what I have," one soldier told me. "I don't want them guys to have to wait twenty years for this."

The problem is that name—induced after-death communication. The problem is the way we respond to topics with any paranormal association at all. There could be other applications for Botkin's method. The visions he elicits might even be used, like lucid dreams, to rehearse for real-life tasks. But in promoting his therapy, Botkin may get his sternest challenge from fellow practitioners of EMDR—people who know how hard it is to gain acceptance for an idea associated with the fringe. In the course of my research, in fact, I called psychologist Dr. Bessel van der Kolk, one of EMDR's leading champions. We had a great discussion, until I brought up Botkin's method.

"That sounds like the patients are making reconstructions after the fact," he said. "No one reports experiences as dramatic as that, and it just sounds . . . too flaky. We've been dismissed, EMDR, in the psychological community for too long. We're getting over that now because people see it works. And nobody wants to go back, to being seen as flaky."

10

You Can't Go Home Again

Has Anyone Else Encountered a Ghost in My Old House?

I never worry about being driven to drink; I just worry about being driven home.

—W. C. Fields

There are a couple of loose threads left dangling here, and for me they are interrelated: What happened in my childhood home, and what happened in Hawaii?

As for the Family Ghost, some readers no doubt feel certain we experienced a *real* spirit—a poltergeist that made its presence known in, shall we say, the *ghostly way*, by banging around at night. Skeptics likely consider my family victims of a collective delusion. I wanted to find out everything I could, so I wrote letters to the family who lived in the house immediately after us and also to the current owners. I'm not going to name any of them, for reasons that will become clear shortly. But I succeeded, first, in contacting the family who moved in right after we left.

My initial conversation, with the mother, was awkward. I started by writing her a letter, telling her that I was interested in her experiences in the house. I made no mention of anything paranormal because I didn't want to risk tainting her response. But this left me at something of a disadvantage on the phone. After all, she didn't know what I really wanted to find out. "So . . . ," I asked her. "How did you like the house?"

"It was fine," she said. "You know . . . nice house."

We made more small talk, comprised mostly of painful silence, till finally she said, "Um, is that all you want?"

"Well," I said, "I was wondering if anything . . . *odd* might have happened there."

"Oh jeez," she said. "I was wondering if you were gonna bring *that* up! That place was *haunted!*"

In the ensuing weeks, I spoke to her, her husband, and two of their daughters. I learned that the girls were supposed to sleep in my sisters' old room but almost never did. "I hated even walking past that room," one of the daughters told me. "If I went in there, I just got creeped out."

The youngest daughter claimed she once saw a stack of pennies just shoot across the room. But the spookiest stuff in this story belongs to the parents. The feature I found most compelling was the father's account: the heavy wooden door he installed in the downstairs bedroom sometimes *boomed* at night, like someone was pounding on it. "I'd see it coming up in the frame," he said, "like it was gonna bust."

He would get out of bed and open the door. But no one was there. Over time, he heard it often enough that he devised a plan to leap from the bed and fling open the door, mid-thump. And the next time it happened, he did. But again, there was no one there. Furious, he lit off through the house, looking for his kids. He admits now, they were way too small to bow the door inward so violently. But when something strange happens in a person's house they often *do* look for prosaic explanations first. As this man stalked through the hall and out into the rest of the house, however, he found his kids asleep upstairs. Out cold.

Does this story mean my old house was haunted by some sort of spirit that went into hiding when we chased it away and came back when a new family arrived? That two fantasy-prone families occupied the house? Or something else, something we haven't thought of yet?

I'm willing to let you decide. But I kept looking. By the time I'd heard from this second family, they had long since moved from my old house. And I'd written a pair of letters to the current occupants.

The first letter just indicated my interest in speaking to them about the house. In the second letter, I was more direct—explaining that we'd experienced some odd happenings there, which my family attributed to a ghost. The owner never responded—who could blame him?—and eventually I knocked on his door, figuring my request for an audience would be much harder to dismiss, face to face on the front porch.

My father came with me because he wanted to see the old house. And I wanted him to see it, too. We had departed almost twenty-five years earlier, and since then we'd lost three immediate family members and added six new babies. So there was a lot of nostalgia involved in going back.

When I knocked on the door, the owner answered—a big man, with thick forearms and broad shoulders. When I announced who I was, he immediately came out onto the porch, waving another man to follow behind him. I felt no threat at all. And once he was outside, with the door closed, he spoke.

"I got your letters," he said, "and I didn't answer because of my son. I don't want to scare him."

It is for this reason that I am not naming the current owner of the house, or the intermediate owners, and I haven't mentioned the location of the house in this book. I suppose, for those interested in sleuthing, all of this would be easy enough to find. But I ask, in consideration of the current owner's wishes, that you not. Besides, there was nothing to see or hear there, anyway, at least as it relates to the Family Ghost. "I think you give that kind of stuff power when you talk about it," the man said. "So I wouldn't talk about it with you. But, ah, nothing's happening."

He quickly turned away from my gaze, and I am perhaps not alone in taking this as a mixed message. My own spidey sense, built up in my years as a journalist, told me that this man had more to say. But it was also clear to me that he wasn't going to share it. This doesn't mean he was hiding some ghostly goings-on. He might merely have wanted to avoid scaring his children with a ghost story. But he was definitely withholding.

He told us that he was a building contractor and that he had refurbished the house himself. The man he called out with him turned out to be an employee, and he quickly seconded his boss's statement: closing his eyes and twisting his lips into a tight grimace, he looked away, shook his head *no* and muttered, "Nothing's happening."

To say I found his conviction underwhelming would be an understatement. But again this could have meant any number of things. Maybe the employee was just a timid guy, uncomfortable with speaking to a journalist? Standing there, I wished, for a moment, that I had entered into this reporting situation as I had so many others—with some leverage. I'd like to have pressed them a bit, to make sure nothing was happening. But the truth was, I both believed and disbelieved in the Family Ghost—and they had no reason to talk to me at all.

Still, we got to see the house. The new owner brought my father and me inside for a tour. And what we saw amazed us. This new family had taken everything we remembered and refashioned it into something else entirely. They knocked down almost every wall upstairs, converting the three bedrooms—including my sisters' old room—into two. Downstairs, the conversions were only slightly less massive. My mother's kitchen was twice its former size, encompassing what had been a separate dining room. And they had even taken our old dark, scary basement and turned it into a cluttered workshop and weight room. The new owner's wife followed us around, smiling tightly the whole time, as we *ooahed* and *aahed* at our now unrecognizable old home.

We left after maybe a fifteen-minute tour. The owner politely walked us to the door, and that was that. We were silent until we got in the car and started the engine. Then my father was the first to speak.

"That's it, Steve," he said. "Our old house is gone."

The Family Ghost, true to its form, whatever its form—unlikely plumbing problem, unlikely ghost—had fled further inquiry.

To my surprise, however, this particular part of my story still wasn't over. And an answer of a kind did arrive—in a most unexpected form and place.

11

Our Time in Hell

Lucid Dreaming and Its Most Potent Lesson for Science and Spirituality

Compared with what we ought to be, we are only half awake.

—William James, "The Energies of Men"

I lay down, planning to put myself to sleep while meditating. I was midway through my trip to Hawaii, on day five of the workshop. My mind was relaxed and focused on LaBerge's suggestion for a simple mantra: *The next time I'm dreaming, I'll remember to recognize that I'm dreaming.*

I'm not sure how long I kept that mantra going. But I distinctly remember the strange sensation of feeling myself disconnect from my body, like a plug pulled from its socket. Then I began to drift, downward, through the floor of the bedroom. I passed through the wood with a subtle, satisfying friction, expecting to see the first story of the building I slept in—the refrigerator, the chairs lining the wall, the screened exit door. But instead I saw . . . stars.

I was in outer space.

And just like that, I was lucid.

There was no need for a state test, no need to ask any questions at all—and again I felt that sensation of zapping into another reality, of going to sleep in one world and coming to, suddenly, in another.

Points of stars appeared in sharp relief, out in the distance. Strange mists, collections of gas, floated pretty as dandelion seeds in the air. My body felt suddenly solid and real. My muscles tensed slightly as I dropped gracefully through the sky. The sensation of descent was so realistic, I even felt my dream hairs stand up on my dream arms. And after a few moments of enjoying all this, I looked down to see where I was headed.

I could see something in the distance, a squarish platform coming into view, but couldn't quite make it out. I felt hyperconscious now. And I knew that was a good thing. As LaBerge had advised us, the version of reality we create when we dream is incredibly tenuous. Look away from an object, and turn back, and its nature can change entirely. LaBerge has developed some techniques to stabilize the dream. My favorite is the act of rubbing my "dream hands" together, like I'm trying to warm them over a fire.

Why does that work?

Well, remember, the dream is much like waking consciousness, without sensory input from the waking world. Rubbing the dream hands together provides the brain with sensory input *from the dream world;* and those sensations create a greater sense of reality, helping the dream persist. But in this dream, of falling through the sky, the air flowing over my skin kept the dream world stable. There was no need to rub my hands at all. And after a few more seconds, I could finally make out what lay below me: the squarish shape I saw was a giant Monopoly board.

A Monopoly board as big as a small island in the Pacific.

As I continued falling, I could see the content of the squares running along its perimeter. There was no Park Place and no Broadway, no Go to Jail square. In fact, all the squares surrounding the board contained images of a man, and a woman, together, in different silhouetted poses. And in the dead center of the board, there was a hole. I was headed straight for it.

I should stop here, for just a moment, to remark that I now count this dream as among the most incredible experiences of my life. I say this because until I had spent long minutes lucid, inside of a dream, I had no idea the power such an experience could hold. The sensory reality of everything I saw was just one aspect: yes, the colors and shapes and the feel of the air rushing around me were incredibly vivid and tactile. And then there is the carnival-like nature of the dream: because no physical law can constrain, existence becomes a kind of boundless circus—life as a character inside of a sci-fi movie or video game. But more profound than any of this is the feeling of finding myself inside an entirely subjective experience—only to find that this interior world seems to be as big as the universe itself.

I got comfortable and began to experiment. I willed myself to move left, right, back, forward, even up. But then I chose a course of action: LaBerge had advised us to approach our dreams with a curious spirit; and when I had attained lucidity in this dream, I had been dropping downward, toward the Monopoly board below; so I wondered what my subconscious mind had cooked up for me; and with a kind of happy, *banzai* recklessness, I again allowed myself to fall. Immediately, like an elevator cut loose from its cabling, I plummeted downward.

I fell faster than before. And in just seconds, I passed through the hole at the board's center. There was a momentary blackness. The walls of the hole I'd dropped into were all around me. And then I was out. Immediately, I saw a second game board beneath this first one. And again, as I drew near, I could make out the details: the squares on this board were filled with images of machinery—cogs and wheels, gears and crankshafts. I passed through the hole in the center of that board, too, like passing through a highway tunnel with the lights out. And beneath that was a third board. But this one was fundamentally different.

Even from a distance, I could see there was almost nothing written on this board. And after a few seconds, what detail there was became clear: all the squares were empty but one, which contained the illustrated image of a single man in midstride. This man appeared in the upper left-hand corner of the board, and again in the center there was another hole. But between these two features was the thing that got my attention: a single word, printed in big, capitalized letters:

HELL

Instinctively, I panicked. *This hole,* I thought, *is the entrance to hell.* I started flying upward, away from the board. I was raised Catholic, after all, and these concepts still hold a primal sort of power. But then I remembered why I had come here—to explore the dream world, and moreover to explore myself. All the dream images I saw, the dream sounds, were just constructs of my own mind. So whatever hell lay below me, it was one of my own creation. *This is a dream,* I reminded myself. *I'm safe.*

I let myself go again. I began dropping toward the HELL hole and actually tried to interpret a dream—for the first time ever—*while* I was in it. *Does this dream mean,* I thought, interpreting it literally, *that if I stay on my current path, I'm going to hell?*

But another thought occurred to me, just as automatically: *No, there is no hell.*

I changed my attitude. I told myself I should be excited. And by the time I closed in on the darkened opening, I found myself feeling much the same way I do on a rollercoaster—filled with anxiety and the expectation of a coming thrill. I was about to see HELL—and live to tell about it. And a moment later, I was inside the hole. Blackness enveloped me. The only sensory input I received was the dream air rushing around my dream face. Then I saw flames below me, flickering. And I even felt the heat. But at this point, I spoke to my own dream: *No,* I said, *this is just what I expect to see. What am I really here to see?*

And suddenly, in an instant, the flames were gone. In fact, I landed. I had come to a stop. I was face down on the floor of a room, in the same position I had fallen. I lifted my dream body, from the torso up, in a push-up position, to look around. My knowledge of where I was came to me, immediately, in the strange way knowledge often arrives in a dream—as an inexplicable flash of insight.

This, I realized, was my maternal grandfather's bedroom. This surprised me because it hardly even *looked* like my grandfather's bedroom. There was a sink in his room, an odd feature, and this room had a sink, too. But beyond that, there was little resemblance. His room was neat and orderly. This room was cluttered, with racks of clothes and bottles of alcohol. And as I took all this in, I had another thought, which seemed to make everything else fall into place: *Oh,* I said aloud, in the dream world. *This is my foundation.*

A tremendous surge of energy rose within me. I felt like there was something I'd been looking for, something that had never quite been clear to me, and now I had all my answers. If the beginning of the dream, before attaining lucidity, felt like disconnecting from a socket, this was like plugging back in. Every fiber of my being seemed alive with electric current. And that was it. The dream was over.

I opened my eyes. This mental image of my grandfather's room, cluttered and unrecognizable, gave way to the sight of my own well-ordered room in Hawaii. I transitioned from one reality to the next in a seamless instant. I had no sense of "waking up" because I had already been awake and aware the whole time. It was as if someone had just switched the computer program I had been running.

I thought for just one beat, and an interpretation of the dream came to me, which I shared that day for the workshop participants and LaBerge in Hawaii. But I'm not going to write about it here. Because all the interpretation I did afterward felt anticlimactic. Over the next few days, in fact, I tried out different ideas but none ever took. What had mattered was the *experience* of the dream. I felt as if I had really touched my own foundation—and the sensation had been electric.

But LaBerge had something more to say on the subject—the Mad Hatter answering questions I hadn't even asked.

IIIIIIIIIII

THE WHOLE TIME I was in Hawaii, I was curious to suss out LaBerge's take on paranormal topics. But for much of the workshop, he seemed to let opportunities to *go there* go sailing by. I wondered if he had made a kind of tactical decision. After all, the acceptance of lucid dreaming has been hindered by its long association with the paranormal, an association that will likely never go away. Robert Waggoner, for instance, has built a sturdy niche for himself in the community of lucid dreamers by arguing that some of the dream characters we encounter may enjoy an objective, independent existence. By way of contrast, LaBerge's own writings on lucid dreaming continue to stand out for their comparatively materialistic scientific outlook.

I was surprised, then, by a lecture he gave on the topic of telepathy. "The question comes up a lot," said LaBerge, "because people report these kinds of strange experiences, both in and outside of dreams, in which they seem to gather information outside the normal sensory channels."

LaBerge, however, had something more than anecdote to share. His story was about a scientific experiment.

LaBerge had been asked, he told us, by a branch of the American military, to study whether people could gain accurate information about the waking world in lucid dreams. And so he helped to construct an experiment, volunteering to do double-duty as one of the subjects. "I didn't expect to succeed," LaBerge said. "But I figured it was worth a shot. People had been reporting *some* success in these matters, and though I was doubtful, I decided to see for myself."

I won't go into all the details of the experiment here. But dreamers, including LaBerge, were charged with opening up an envelope in a lucid dream, examining its contents, and awaking to write a detailed report. In each case, the dream envelope they were supposed to open had a real-life corollary. And the goal of the experiment was both simple and wild: Would the contents of the sealed, waking-life envelope, when opened, match the contents of the envelope opened *in dreams?*

According to the experiment's protocol, each dreamer had a 20 percent chance of scoring a hit, purely by luck. But the hit rate produced by the experiments was 33 percent. Further, dreamers were more than three times as likely to score a close match with the target image than any *non*-target image. "I have to admit I was greatly surprised by these results," said LaBerge. "But the statistics seem to demonstrate that *something* is happening here."

LaBerge also showed us a particularly evocative example of success from his own archives of that experiment. Upon awakening from his own lucid dream, he drew a series of wavy lines and four words: *sandy, patchy, dark, light.*

When the sealed envelope was opened later, without LaBerge's presence, the target image turned out to be a set of rolling sand dunes, with heavily shaded areas offset by patches of bright sunlight.

As LaBerge acknowledged, any one example of success could be dismissed as coincidence. Maybe even this spot-on match was just that. But the experiment he took part in rules out coincidence as any sort of explanation for the results as a whole. "We found compelling results," said LaBerge, "that were, statistically, significantly above chance. But the results we got were similar to other credible research, in that whatever's going on here can't really be operationalized. There is some other way of knowing then our standard senses, it seems, but it's a weak signal—and hard to pick up on."

After that lecture, I caught up to LaBerge before he could leave and asked him what we were to make of this. Like anyone else, I'd had experiences of my own that could have been put down to prophecy—or more likely, mere coincidence. One woman in class even shared a story about dreaming of a fatal accident that befell a young boy *before* it happened. How should someone view a potentially telepathic experience when it could just as easily have been random chance?

"I'd say," said LaBerge, "that if you believe you have picked up on information in this way, the appropriate response is to *believe* what the experience is telling you. In other words, let it open your mind to the fact that there is more to this world than we know."

And so, over time, it was clear to me that LaBerge was more complicated than I had known; he even told us of a lucid dream he had in which he asked to see his highest self and flew, in his dream car, to a place he could only describe as a "holy nothing, a beautiful nothing, an ecstatic nothing."

His dream mirrored the classic peak meditative experience, which he described as "the total absence of any sense of self, and a sudden, joyful connection to everyone and everything."

Over time, said LaBerge, he had come to see the dream as, perhaps, a "metaphor for what happens when we die. When the drop of water, as it were, realizes its true identity as part of the ocean."

At another juncture, LaBerge sat by quietly as his assistant and friend, Keelin, shared an incredible story about healing herself in a lucid dream. She was struggling, in real life, with excessive menstrual bleeding. Her doctor was advising her to have a hysterectomy. Your *uterus* is tilted, she was told. But after this diagnosis, Keelin habitually misspoke, telling her friends: "My *universe* is tilted."

She didn't want to have a hysterectomy, but she feared she would have no choice. And she was reaching the stage at which circumstances would force a decision. Then she had a lucid dream in which she was able to reach directly inside her body and heal herself. It was an energizing dream that made her feel as if she had personally done everything she could to avoid the hysterectomy. But the crazy thing was, after the dream, her excessive bleeding stopped. Immediately. And it never came back.

As she put it to me, "That's not science, I understand, but—"

Yes, but . . . what shall we do?

I mean, we live in a world where we *have* to define this sort of thing, don't we, as nonsense or reality? If an event like this isn't science, shouldn't it be disregarded entirely?

The answer LaBerge's entire career suggests is, in a word, *no.* We don't have to make the choice that popular culture gives us; we don't have to choose one and dispense with the other. This is not a world of binary opposites. We just live that way. We could, in fact, choose to believe what Keelin's experience is telling us: that there is more to this world than we know. And then, it seems, the most rational response might be to *explore* it—to see if the events she described could really be so.

The problem is, in this culture, when a claim carries the whiff of hoo ha, too much of the intelligentsia goes running for the hills. And on the opposite pole, when a scientific finding seems to undercut our spiritual belief, we dismiss it—that is, if we even bother to read it. "We live in a world in which there are a lot of extremes offered," LaBerge said, summing up the current culture wars. "The view of many modern neuroscientists is that the subjective experience of consciousness, is what they call an epiphenomenon, or product, of brain function, and therefore scientifically unimportant. But then I've also met Tibetan Buddhists who think the brain's only function is to keep our ears apart."

We all laughed. But LaBerge kept going.

"I recommend," he said, "*balance* in these matters."

The thing is, balance can be so hard to come by.

To the media, balance is best maintained by gathering the most extreme combatants, wielding the most far-flung opinions, to debate. The result, nearly always, is a shrill argument in which the truth seems to lie, tantalizing and undiscovered, somewhere between two self-serving accounts. And among the most strident of believers and atheists, there is no balance at all—only right and wrong, the heretics and the saved, the intelligent and the foolish.

LaBerge captured this dynamic most thoroughly, inadvertently, with the aid of a prop.

Standing up from his chair in the front of the room, he retrieved a mask bearing the face of the devil: big fat tongue, blood red skin, menacing expression—the whole deal. LaBerge held it to himself tightly and smiled. This tableau was already dreamlike, the barefoot scientist and the devil, but he was just getting started.

He told us the story of one of his own dreams, in which he confronted an ogre. He was lucid, and he knew the creature could not harm him. But still, he backed away. The beast was smelly, so rank he could barely go near it.

From where, he wondered, *in the recesses of my own self, did this awful creature emerge?*

This question was the source of his curiosity, the thought that enabled him to get past his own reluctance and advance toward the ogre. The smell, at first, was still overpowering. But he forced himself up closer to the animal. And he resolved to accept this ogre, to accept its ugliness and its odor as part of himself. And magically, as in the way of dreams, the ogre simply *melted into him*. LaBerge awoke then and felt—great.

"Dreams are often about this sense of feeling fragmented," LaBerge said, "about taking the parts of ourselves we consciously hide and integrating them. Such dreams are about acceptance."

In this dream, the things LaBerge hides had manifested in the noxious form of an ogre. I was surprised when LaBerge, as he finished this story, suddenly shifted his gaze toward me. But then he explained how his dream revealed the substance of my own dream, in which I dropped through a series of Monopoly boards on my way to hell.

The holes I passed through, he said, those symbolized the concept of wholeness, of integration. And as for HELL, the word that stood between the little man on my last board and the wholeness waiting for him in the center?

"If you take this interpretation," said LaBerge, "hell is this sensation of not feeling integrated; hell is this sense of not feeling whole."

In this view, the act at the end, of *accepting* my foundation, was like LaBerge absorbing his ogre. No wonder, when I did it, I felt plugged back in.

A lot became clear to me then. Because LaBerge's dream analysis does a pretty neat job, not only of explaining one function of dreaming as a tool for psychological healing, but also as a metaphor, explaining our society's current relationship to the paranormal.

We are not integrated. We are carved into tribes, believers and unbelievers. And as a society, I think, we will persist in a kind of hell—a hell of separateness—if we do not understand what it is we've done to each other.

The saved and the damned.

Muslims and infidels.

Brights and dulls.

The rationalists and materialists, lined up against the dark forces of "superstition."

And where do these divisions get us? No closer to one another, certainly, and no closer to any real answers.

Instead, we demonize each other, putting the equivalent of devil masks on our foes. But in the case of science and spirituality, these seeming opposites might yet converge. We can even *see* them come together, in Andrew Newberg or Stephen LaBerge—two researchers who used the tools of modern science to verify the experiences mystics had been reporting for millennia. And in each case what they found as a guiding principle was not division—but *connection*.

The question is whether or not we're prepared to accept these findings, to accept a world in which religion and science don't have to clutch at each other's throats. The question is whether or not we're prepared to accept a world in which science and spirituality really do serve each other.

The Big Ghost, of course, hovering at the back of this entire discussion, is God. Human beings are always fighting about which version of God to worship, or whether any God exists at all. But it seems to me we are only likely to find answers about the nature of the universe, or the possibility of a creator, *if we look*. And we can't do that in any meaningful way if our only commitment is to the answers we've presupposed.

At this book's outset, I mentioned the noted atheist Sam Harris, whose words I find instructive here, as they speak to the kind of opportunity we now hold, in the advancement of science, and the contribution it can make to human spirituality.

"For millennia," writes Harris, "contemplatives have known that ordinary people can divest themselves of the feeling that they call 'I' and thereby relinquish the sense that they are separate from the rest of the universe. This phenomenon, which has been reported by practitioners in many spiritual traditions, is supported by a wealth of evidence—neuroscientific, philosophical, and introspective. Such experiences are 'spiritual' or 'mystical,' for want of better words, in that they are relatively rare (unnecessarily so), significant (in that they uncover genuine facts about the world), and personally transformative. They also reveal a far deeper connection between ourselves and the rest of the universe than is suggested by the ordinary confines of our subjectivity. . . . A truly rational approach to this dimension of our lives would allow us to explore the heights of our subjectivity with an open mind, while shedding the provincialism and dogmatism of our religious traditions in favor of free and rigorous inquiry."

I'd add to this that such an approach is only possible if the materialists among us let go of their own dogma, too, and stop snickering at the paranormal, stop snickering at the possibility that God, or something like Him, might crop up in this search. I realize how difficult it is to make these kinds of shifts—for Pat Robertson to start thinking of his Bible as a collection of books in need of further editing. Or for Richard Dawkins to act on his intellectual understanding—that science might one day yield up evidence of all he can't believe. For some, these shifts may prove too great. But the opportunity lies before us just the same. The war is over, if we want it, and what we get in return is each other. What we get in return is the thrill of a "free and rigorous" inquiry into the true nature of reality—and what it means to be human.

When my time in Hawaii was over, and LaBerge's final lecture ended, the devil mask lay out in full view, for everyone to see—and a party started.

The workshop participants slowly passed around beer and wine. And I ultimately joined them. But before I did, I walked outside, over the grass and into the dark.

The night sky at Kalani was forever painted with stars—magisterial and immense—and I lingered there for a time. And, I must confess, I thought about the Family Ghost.

That strange family story, which had come to be a mark of embarrassment for me in some circles, had set me on this search. But standing there, seeing every distant point of light look so incredibly near, I realized that I no longer cared if I had been set on this course by a fateful spirit—or faulty plumbing. Because down through the millennia, from the first time one of our ancestors started looking for answers in the movement of the stars, or the cracks in a turtle's shell, whatever it is that we call paranormal, whatever it is that so confounds us—real or imagined, measurable or not—isn't banging on the roof and walls, trying to get in. Whatever has been making all this noise, is already here. And as I'd learned from my own search, there are ways we can access all this for ourselves—to uncover whatever truths we might find, to enjoy whatever peace we can attain.

I sat in the grass for a while, watching the stars, and enjoyed my own private celebration. Then I went upstairs, to the party that had been going on without me.

ACKNOWLEDGMENTS

After having spent many years working on this book, in one form or another, I have a lot of people to thank.

My agent, Andrew Stuart, for all his support and advice.

My editors, Eric Brandt, who handled this book at its inception, and especially Roger Freet, who steered it home with wit and an evident passion to serve both me and, more importantly, the reading public. Thanks for making this book better. Also Amanda Wood, Christina Bailly, and Darcy Cohan for all their help.

I'd also like to thank all my sources for this book, from those who shall remain nameless to those who gave so generously of their time to make this book possible. I'd like to single out Dianne Gray, Willoughby Britton, Stuart Hameroff, Jim Trolinger, Jack Tuszynski, Angelia Joiner, Dean Radin, Jessica Utts, Edgar Mitchell, Andrew Newberg, Mark Robert Waldman, Keelin, Stephen LaBerge and Allan Botkin for reading over select portions of this manuscript.

My colleagues in Philadelphia and elsewhere, who have contributed to my career in various ways: Theresa Conroy, Will Bunch, Stu Bykofsky, Sasha Issenberg, Benjamin Wallace, Jessica Pressler, Lee Gutkind, Liz Spikol, Brian Hickey, Brian McManus, Jeff Barg, Chris McDougall, Ralph Cipriano, and Kia Gregory. Of course, those with whom I have shared the most beers and long conversations belong in a special category in my heart, mind and liver: Chris Brennan, Mike Newall, Dan Kennedy, Rich Rys, Jason Fagone, Jeff Deeney, Chris Thompson, Jonathan Valania and Michael Rosenwald.

To all those who have ever hired me to work in journalism, including Sara Kelly, Tim Whitaker, Larry Platt, and Tom McGrath, you have my eternal gratitude.

To Tim Haas, for his help with nettlesome technology.

Stephen H. Segal belongs in his own category, for all his help in the writing and thinking that went into this book. So does Lou Gentile, who got me started on this book and unfortunately did not live to see it. Thank you both for your friendship.

To my family, both blood and in-laws, for all the good times, advice, support and love; for the days that have already been and the days yet to come.

To Lisa Stachler Volk, my wife, for the best years of my life and a honeymoon that will never end.

And my father, Gerald Volk, for teaching me that "Life is good."

NOTES AND SOURCES

A note from the author: Sources are listed in order of appearance within each chapter. In instances where the source is found online, I listed the most recent date that I accessed the site in question. Personal interviews conducted by the author are simply listed as "interview."

INTRODUCTION

Richard Dawkins, "Time to Stand Up," *A Devil's Chaplain,* (Mariner Books, 2003): 156–161.

Interview, Christopher Hitchens, Martin Amis, with Jeffrey Goldberg, accessed October 24, 2010, http://www.theatlantic.com/culture/archive/2010/08/hitchens-talks-to-goldberg-about-cancer-and-god/61072/

Rodney King, May 1, 1992, accessed October 24, 2010, http://www.youtube.com/watch?v=tgiR04ey7-M

Author's note: The Lou Gentile material comes from reporting I did in June–July 2006.

"Ghost Sightings Highest in 25 years," no author given, *Telegraph,* April 26, 2010, accessed October 25, 2010, http://www.telegraph.co.uk/news/newstopics/howaboutthat/7631387/Ghost-sightings-highest-in-25-years.html

Michael Sheridan, "Winston Churchill, Dwight Eisenhower Covered Up UFO Sighting in England, Letter Claims," *New York Daily News*, August 5, 2010, accessed October 25, 2010, http://www.nydailynews.com/news/world/2010/08/05/2010–08–05_winston_churchill_dwight_d_eisenhower_covered_up_ufo_sighting_in_england_letter_.html

Nigel Watson, " 'UFO Hacker' Tells What He Found," *Wired*, June 21, 2006, accessed August 6, 2010, http://www.wired.com/techbiz/it/news/2006/06/71182

Joy Basi, "Taxi Drivers and Ghost," *Solomon Times*, October 14, 2008.

http://www.merriam-webster.com/dictionary/supernatural

http://dictionary.reference.com/browse/paranormal

http://www.merriam-webster.com/medical/paranormal

http://unabridged.merriam-webster.com/cgi-bin/unabridged?va=paranormal&x=15&y=4

Irving Kirsch, "Specifying Nonspecifics: Psychological Mechanisms of Placebo Effects," in *The Placebo Effect: An Interdisciplinary Exploration* (Harvard Univ. Press, 1997): 166–80.

Margaret Kemeny et al., "Placebo Response in Asthma: A Robust and Objective Phenomenon," *Journal of Allergy and Clinical Immunology* 119, no. 6 (June 2007): 1375–81.

Steve Silberman, "Placebos Are Getting More Effective. Drugmakers Are Desperate to Know Why," *Wired*, August 24, 2009, accessed October 25, 2010, http://www.wired.com/medtech/drugs/magazine/17–09/ff_placebo_effect

Robert Carroll, "hypnosis," *The Skeptic's Dictionary*, accessed October 25, 2010, http://www.skepdic.com/hypnosis.html

Laurence Armand French, "The False Memory Syndrome: Clinical/Legal Issues for the Prosecution," *Journal of Police and Criminal Psychology* 11, no. 2 (1996): 38–41. (For anyone who is interested, this French article is a brief, solid, and far more balanced primer than *The Skeptic's Dictionary* on the difficulties in assessing recovered memories.)

A. M. Cyna, "Hypnosis for Pain Relief in Labour and Childbirth: A Systematic Review," *British Journal of Anaesthesia* 93, no. 4 (2004): 505–11.

Steven Gurgevich, "Clinical Hypnosis and Surgery," *Alternative Medicine Alert* 6, no. 10 (October 2003): 109–20.

Guy H. Montgomery et al., "The Effectiveness of Adjunctive Hypnosis with Surgical Patients: A Meta-Analysis," *Anesthesia and Analgesia* 94, no. 6 (June 2002): 1639–45.

Jeff Hughes, "Occultism and the Atom: The Curious Story of Isotopes," *Physics World* (September 2003): 31–35.

David Millett, "Hans Berger: From Psychic Energy to the EEG," *Perspectives in Biology and Medicine* 44, no. 4 (Fall 2001): 522–42.

I recommend a couple of articles I found on the incredible skepticism leveled at the inventions of the lightbulb and the airplane—both of which in their day were treated almost as harshly as paranormal claims.

A. Gelyi, "A Short History of Incandescence Lamps," *Telegraphic Journal and Electrical Review* (February 14, 1885): 139–40.

Simon Newcomb, "Is the Airship Coming?" *McClure's* 17, no. 5 (September 1901): 432–35.

Plato, *The Republic*, ed. C. D. C. Reeve (Hackett, 2005): 297–326.

Ward Hill Lamon, *Recollections of Abraham Lincoln*, 1847–1865, ed. Dorothy Lamon Teillard (Univ. of Nebraska Press, 1911): 114–18.

Author's note: For a skeptical take on Lincoln's seeing it coming, see Joe Nickell, "Paranormal Lincoln," *Skeptical Inquirer* 23, no. 3 (May/June 1999), accessed August 7, 2010, http://www.csicop.org/si/show/paranormal_lincoln/

Colin Ross et al., "Paranormal Experiences in the General Population," *Journal of Nervous and Mental Disease* 180, no. 11 (1992): 357–61.

Angela Joiner, "Possible UFO Sighting," *Stephenville Empire-Tribune*, January 10, 2008, p. 1.

Skip Hollandsworth, "The Searcher," *Texas Monthly* (April 2008); accessed October 26, 2010, http://www.texasmonthly.com/preview/2008–04–01/letterfromstephenville

John Horgan, *The End of Science: Facing the Limits of Knowledge in the Twilight of the Scientific Age* (Broadway Books, 1997).

"What Is the Universe Made Of?" *Universe 101, Our Universe*, NASA web site, accessed October 26, 2010, http://map.gsfc.nasa.gov/universe/uni_matter.html

Avshalom Elitzur, ed., *Quo Vadis Quantum Mechanics* (Springer, 2005): 73–82.

Karl Pribram, *Languages of the Brain* (Brooks/Cole, 1977).

Author's note: Pribram has focused on the hologram as an explanation for human consciousness, particularly memory storage.

David Bohm, *Wholeness and the Implicate Order* (Routledge Classics, 1980).

Albert Einstein, personal letter, quoted by Freeman Dyson in *Disturbing the Universe* (Basic Books, 1981), 187–93.

Brian Greene, *Fabric of the Cosmos* (Vintage, 2004), 127–42.

Zeeya Merali, "Back from the Future," *Discover*, accessed October 26, 2010, http://discovermagazine.com/2010/apr/01-back-from-the-future

For further reading: A full selection of Tollaksen's work is available online, accessed September 1, 2010, http://arxiv.org/find/all/1/all:+tollaksen/0/1/0/all/0/1

Author's note: The following selection of articles and book references is intended to provide readers with a fairly comprehensive overview of the roles the amygdala and other brain structures play in the automatic processing of information and also in the construction and defense of our beliefs. I incorporated interview material with Dr. Andrew Newberg (covered in chapter 7).

M. P. Ewbank, "The Interaction Between Gaze and Facial Expression in the Amygdala and Extended Amygdala Is Modulated by Anxiety," *Frontiers in Human Neuroscience* (July 7, 2010), accessed October 26, 2010, http://www.ncbi.nlm.nih.gov/pmc/articles/P.M.C2906373/

H. J. van Marle et al., "Enhanced Resting-State Connectivity of Amygdala in the Immediate Aftermath of Acute Psychological Stress," *Neuroimage* 53, no. 1 (October 2010): 348–54.

L. M. Shin, "The Neurocircuitry of Fear, Stress, and Anxiety Disorders," *Neuropsychopharmacology* 35, no. 1 (January 2010): 169–91.

M. Browning et al., "The Modification of Attentional Bias to Emotional Information: A Review of the Techniques, Mechanisms, and Relevance to Emotional Disorders," *Cognitive Affective Behavioral Neuroscience* 10, no. 1 (2010): 8–20.

T. Lidaka, "Forming a Negative Impression of Another Person Correlates with Activation in Medial Prefrontal Cortex and Amygdala," *Social Cognitive and Affective Neuroscience* (August. 6, 2010): DOI: 10.1093/scan/nsq0722010.

S. Wiethoff, "Response and Habituation of the Amygdala During Processing of Emotional Prosody," *Neuroreport* 20, no. 15 (October 7, 2009): 1356–60.

D. Wildgruber et al., "Cerebral Processing of Linguistic and Emotional Prosody: fMRI Studies," *Progress in Brain Research* 156 (2006): 249–68.

A. Marchewka et al., "Grey-Matter Differences Related to True and False Recognition of Emotionally Charged Stimuli—A Voxel Based Morphometry Study," *Neurobiology of Learning and Memory* 92, no. 1 (July 2009): 99–105.

A. J. Calder et al., "Neuropsychology of Fear and Loathing," *Nature Reviews Neuroscience* 2 (May 2001): 352–63.

N. O. Rule et al., "Voting Behavior Is Reflected in Amygdala Response Across Cultures," *Social Cognitive and Affective Neuroscience* 5, nos. 2–3 (June 2010): 349–55.

J. B. Freeman et al., "The Neural Origins of Superficial and Individuated Judgments About Ingroup and Outgroup Members," *Human Brain Mapping* 31, no. 1 (January 2010): 150–59.

M. Deppe et al., "Evidence for a Neural Correlate of a Framing Effect: Bias-Specific Activity in the Ventromedial Prefrontal Cortex During Credibility Judgments," *Brain Research Bulletin* 67, no. 5 (November 15, 2005): 413–21.

C. M. Funk et al., "The Functional Brain Architecture of Human Morality," *Current Opinion in Neurobiology*, 19, no. 6 (2009): 678–81.

H. Takahashi et al., "Neural Correlates of Human Virtue Judgment," *Cerebral Cortex* 18, no. 8 (August 2008): 1886–91.

L. Young et al., "Investigating Emotion in Moral Cognition: A Review of Evidence from Functional Neuroimaging and Neuropsychology," *British Medical Bulletin* 84 (2007): 69–79.

J. B. Peterson et al., "Complexity Management Theory: Motivation for Ideological Rigidity and Social Conflict," *Cortex* 38, no. 3 (June 2002): 429–58.

C. K. De Dreu et al., "Mental Set and Creative Thought in Social Conflict: Threat Rigidity Versus Motivated Focus," *Journal of Personality and Social Psychology* 95, no. 3 (September 2008): 648–61.

T. A. Hare et al., "Contributions of Amygdala and Striatal Activity in Emotion Regulation," *Biological Psychiatry* 57, no. 6 (March 15, 2005): 624–32.

David H. Zald et al., "The Human Amygdala and the Emotional Evaluation of Sensory Stimuli," *Brain Research Reviews* 41, no. 1 (January 2003): 88–123.

W. C. Drevets, "Reciprocal Suppression of Regional Cerebral Blood Flow During Emotional Versus Higher Cognitive Processes: Implications for Interactions

Between Emotion and Cognition," *Cognition and Emotion* 12 (1998): 353–85.

Alok Jha, "Where Belief Is Born," *Guardian,* June 30, 2005, accessed http://www.guardian.co.uk/science/2005/jun/30/psychology.neuroscience

Kathleen Taylor, *The Science of Thought Control* (Oxford Univ. Press, 1996), 127–45.

These last two books are excellent primers:

Milton Rokeach, *The Open and Closed Mind* (Basic Books, 1960).

Charles Hampden-Turner, *Maps of the Mind* (Collier, 1982).

The Impact of Emotion in the American Public's Assessments of and Reactions to Terrorism, research brief published by the Institute for Homeland Security Solutions, June 2010, accessed October 26, 2010, https://www.ihssnc.org/portals/0/ . . . /IHSS_Research%20Brief_Singer.pdf

Ara Norenzayan et al., "Mortality Salience and Religion: Divergent Effects on the Defense of Cultural Worldviews for the Religious and the Non-Religious," *European Journal of Social Psychology* 39 (2009): 101–13.

Jean Faber, "Information Processing in Brain Microtubules," presented at Quantum Mind Conference (2003), accessed October 26, 2010, qubit.lncc.br/files/jfaber_InfProc.MT.pdf

Shi Chunhua, "Quantum Information Processing in the Wall of Cytoskeletal Microtubules," *Journal of Biological Physics* 32, no. 5 (November 2006): 413–20.

T. J. Craddock, "Information Processing Mechanisms in Microtubules at Physiological Temperature: Model Predictions for Experimental Tests," *Biosystems* 97, no. 1 (July 2009): 28–34.

Karl Pribram, *Rethinking Neural Networks* (Lawrence Erlbaum, 1993): 216, 324–26.

T. J. Kaptchuk et al., "Components of Placebo Effect: Randomized Controlled Trial in Patients with Irritable Bowel Syndrome," *British Medical Journal* 336 (2008): 999–1003.

Thomas Kuhn, *The Structure of Scientific Revolutions,* 3rd ed. (Univ. of Chicago Press, 1996). *Author's note:* Seminal and eminently readable.

Tener Edis, "Quantum Magic," *Secular Outpost,* accessed October 26, 2010, http://secularoutpost.infidels.org/2007/11/quantum-magic.html

Michel Shermer, "Quantum Quackery," *Scientific American* (January 2005), accessed October 26, 2010, http://www.scientificamerican.com/article.cfm?id=quantum-quackery

Mark Buchanan, "Do Birds See with Quantum Eyes?" *New Scientist* (May 2008), accessed October 26, 2010, http://www.newscientist.com/article/mg19826544.000-do-birds-see-with-quantum-eyes.html

"Quantum Biology Has Come In from the Cold," editorial, *New Scientist* (February 2010), accessed October 26, 2010, http://www.newscientist.com/article/mg20527462.500-quantum-biology-has-come-in-from-the-cold.html

Gregory S. Engel et al., "Evidence for Wavelike Energy Transfer Through Quantum Coherence in Photosynthetic Systems," *Nature* 446 (April 12, 2007): 782–86.

A. Zeilinger et al., "Quantum Interference Experiments with Large Molecules," *American Journal of Physics* 71, no. 4 (2003): 319–25.

Blake Wilson, "Stray Questions for David Eagleman," *Paper Cuts*, *New York Times* blog, July 10, 2009, accessed October 26, 2010, http://papercuts.blogs.nytimes.com/2009/07/10/stray-questions-for-david-eagleman/

Hal Arkowitz, "Why Science Tells Us Not to Rely on Eyewitness Accounts," *Scientific American* (January 2010), accessed October 26, 2010, http://www.scientificamerican.com/article.cfm?id=do-the-eyes-have-it

The following sources provide a good overview of some of the more contentious, even strident claims of the New Atheists, including Dawkins's support of the idea that atheists be dubbed "brights." This list, however, is not even close to comprehensive. The Eagleman lecture is provided for contrast.

Richard Dawkins, *Atheism and Faith,* http://videosift.com/video/Richard-Dawkins-Atheism-and-Faith and http://wn.com/I%27m_an_atheist,_BUT_____by_Richard_Dawkins_1_of_6

"Richard Holloway in Conversation with Richard Dawkins" accessed October 26, 2010 (April 2008) http://www.stcuthbertscolinton.org.uk/wordweb/conversation.htm

David Eagleman, "On Uncertainty," accessed October 26, 2010, http://www.vimeo.com/12543623

Richard Dawkins, "The Future Looks Bright," *Guardian*, June 21, 2003, accessed October 26, 2010, http://www.guardian.co.uk/books/2003/jun/21/society.richarddawkins. *Author's note:* The New Atheists do not all agree with Dawkins's endorsement of the "brights" idea. Christopher Hitchens has taken very public exception to it.

Christopher Hitchens, *Letters to a Young Contrarian* (Basic Books, 2001): 55.

Sam Harris, "Is Religion Built Upon Lies?" online dialogue between Harris and Andrew Sullivan, accessed October 21, 2010, http://www.beliefnet.com/Faiths/Secular-Philosophies/Is-Religion-Built-Upon-Lies.aspx?p=3

Author's note: These books provide a firm grounding in attempts to integrate science and spirituality.

Jean Gebser, *The Ever-Present Origin* (Ohio State Univ. Press, 1986).

Ken Wilber, *Up from Eden* (Quest Books, 1996).

Ken Wilber, *A Brief History of Everything* (Shambhala, 2001).

CHAPTER 1: ON DEATH AND NOT DYING

Ernest Hemingway, "A Clean Well-Lighted Place," *The Complete Short Stories of Ernest Hemingway: The Finca Vigia Edition* (Scribner, 1998), 288–91.

Nick Cave, "*Dig, Lazarus, Dig!*" CD (Mute, 2009).

Fern Welch, Interview, March 2009.

Christopher Reed, "Psychiatrist Who Identified Five Stages of Dying—Denial, Anger, Bargaining, Depression and Acceptance," *Guardian*, August 31, 2004.

Elisabeth Kübler-Ross, *On Death and Dying* (Scribner Classics, 1997), 35–36 (story of her beginning work in death studies at Chicago hospital), 172–74 (story of farmer).

Russell Friedman, "Broken Hearts," *Psychology Today* (September 21, 2009), accessed October 21, 2010, http://www.psychologytoday.com/blog/broken-hearts/200909/no-stages-grief

"The Myth of the Stages of Dying, Death, and Grief," *Skeptic* 14, no. 2 (2008), accessed October 26, 2010, grief.net/Myth%20of%20Stages.pdf

Author's note: It may appear a contradiction that her five stages are often critiqued, yet *ODAD* remains required reading. In the end, academics now take a less dogmatic view of Kübler-Ross's stages—seeing them less as gospel and more as a rough guide. But for a primer on just what's happening inside one of those darkened hospital rooms, she remains tough to beat.

Raymond Moody, *Life after Life* (HarperOne, 2001).

Author's note: The cadre of people gathered around Kübler-Ross's memory act, in metaphorical terms, not altogether unlike soldiers manning a barricade. The people I interviewed for this story, particularly Dianne Gray, all say they encounter a wide variety of people who see Kübler-Ross in diametrically opposed ways—both positive and negative. That Kübler-Ross herself felt personally wounded by the criticism of her that mounted over the years is not surprising; that the people who knew her best are still dealing with and responding to that criticism seems something else entirely and speaks to our desire to build up or tear down people depending mostly on whether or not they seem to symbolize our own worldviews.

Holcomb B. Noble, "Elisabeth Kübler-Ross, 78, Dies; Psychiatrist Revolutionized Care of the Terminally Ill," *New York Times*, August 26, 2004.

Ken Ross, Interviews, March, April, June, and August 2009.

The following nine articles provide a strong overview of the role cognitive dissonance plays in the way we develop and defend our worldviews.

Vincent van Veen, "Neural Activity Predicts Attitude Change in Cognitive Dissonance," *Nature Neuroscience* 12, no. 11 (2009): 1469–74.

C. S. Carter, "Anterior Cingulate Cortex and Conflict Detection: An Update of Theory and Data," *Cognitive, Affective, and Behavioral Neuroscience* 7, no. 4 (2007): 367–79.

V. van Veen, "Conflict and Cognitive Control in the Brain," *Current Directions in Psychological Science* 15, no. 5 (2006): 237–40.

Joshua Gowin, "Why It's Hard to Stop Believing in Santa Claus," *Psychology Today*, (November 17, 2009), accessed October 26, 2010. http://www.psychologytoday.com/blog/you-illuminated/200911/why-it-s-hard-stop-believing-in-santa-claus

Daniel Levine, "Cognitive Dissonance, Halo Effects, and the Self-Esteem Trap," *Psychline* 2, no. 3 (1998): 25–26.

Daniel Levine, "Negotiating Cognitive Dissonance," *Explorations in Common Sense and Common Nonsense*, 123–50, accessed October 26, 2010, http://www.uta.edu/psychology/faculty/levine/EBOOK/

Carol Tavris and Elliot Aronson, *Mistakes Were Made, but Not by Me* (Houghton Mifflin, 2007).

J. A. Bargh et al., "The Unbearable Automaticity of Being," *American Psychologist* 54, (1999): 462–79.

Rose Winters, Interviews, January, February, March, and May 2009.

Elisabeth Kübler-Ross, *The Wheel of Life* (Touchstone, 1997): 22–25 (parents, early childhood), 35–36 (early inclination to act as protector of those weaker than herself), 41–42 (early history with religion), 47–48 (conflict with father over career), 110 (discriminated against for gender), 114–17 (Manhattan State Hospital), 129–34 (her first lecture on death and dying), 176–78 (Mrs. Schwartz), 188 (claims to interview 20,000 people), 201–8 (history with Barham, whom she refers to as "B"). *Author's note*: Like *ODAD*, this book is referenced often. Required reading for anyone interested in Kübler-Ross.

Mwalimu Imara, Interviews, October, November 2009, January, February 2010. *Author's note:* Imara's name when he first worked with Kübler-Ross was Renford Gaines. He subsequently changed his name to reflect his African heritage. For clarity's sake, I refer to him by his current name throughout this book.

Elisabeth Kübler-Ross, *The Tunnel and the Light* (Marlowe, 1999): 86–87.

Loudon Wainright, "A Profound Lesson for the Living," *Life*, November 21, 1969: 36–43.

Author's note: One of the great, confusing aspects of Kübler-Ross's career arises from her flair for exaggeration. She found skeptics such a drag that she sought to quell their objections by claiming to study twenty thousand people who had died and come back. The math alone suggests these figures were impossible. If she had, for instance, run across one person a day, every day of the week, with an NDE story, it would have taken her fifty-five years to reach twenty thousand. Imara puts this down to "Elisabeth's way of saying she had accumulated plenty of evidence. This wasn't a handful of stories." His claim is that he and Kübler-Ross filled two deep filing cabinet drawers with such stories. In short, Kübler-Ross accumulated enough stories to qualify as "plenty," but far less than twenty thousand.

For the 20,000 reference, see Elisabeth Kübler-Ross, "Living and Dying," *On Life After Death,* (1991), chapter 2.

American Heart Association, "History of CPR," accessed October 26, 2010, http://www.americanheart.org/presenter.jhtml?identifier=3012990

G. R. S. Mead, *The Vision of Aridaeus* (Kessinger, 2010): 17–25.

Thomas De Quincey, *Confessions of an English Opium-Eater*: 70, accessed October 21, 2010, www2.hn.psu.edu/faculty/jmanis/tdquincey/Opium-Eater.pdf

Carol Zaleski, *The Life of the World to Come* (Oxford Univ. Press, 1996): 70.

Pim van Lommel, "Near Death Experience in Survivors of Cardiac Arrest: A Prospective Study in the Netherlands," *Lancet* 358 (2001): 2039–42.

Sam Parnia et al., "A Qualitative and Quantitative Study of The Incidence, Features and Aetiology of Near Death Experiences in Cardiac Arrest Survivors," *Resuscitation* 48, no. 2 (February 2001): 149–56.

Bruce Greyson, "Incidence and Correlates of Near-Death Experiences in a Cardiac Care Unit," General Hospital Psychiatry 25, no. 4, (July–August 2003): 269–76.

Jeffrey Long, *Evidence of the Afterlife* (HarperOne, 2010): 5–19.

Alex Tsakiris, "Christian Theologian Claims Near Death Experience Not Communications with Divine," *Skeptiko*, July 7, 2010, accessed October 26, 2010, http://www.skeptiko.com/christian-theologian-claims-near-death-experience-not-devine/

Russel Noyes Jr., "Aftereffects of Pleasurable Western Adult Near Death Experiences," *Handbook of Near Death Experiences* (Praeger, 2009): 41–62.

Diane Corcoran, Interview, October 2009.

Glen Brimer, Interview, October 2009.

Mary Roach, *Spook* (Norton, 2006).

Alex Tsakiris, "Dr. Jeffrey Long Takes on Critics of *Evidence of the Afterlife*", *Skeptiko,* accessed October 26, 2010, http://www.skeptiko.com/jeffrey_long_takes_on_critics_of_evidence_of_the_afterlife/#more–649

Sam Parnia, Interview, January 2010.

D. Luke, "Lecture report: Inducing Near-Death States Through the Use of Chemicals—Dr. Ornella Corazza," *Paranormal Review* 43 (2007): 28–29.

Peter Fenwick, "Science and Spirituality: A Challenge for the 21st Century," talk delivered at the Bruce Greyson Lecture from the International Association for Near-Death Studies 2004 Annual Conference, accessed October 26, 2010, www.larslanke.nl/download/Science%20and%20Spirituality.pdf

Christopher C. French, "Near-Death Experiences in Cardiac Arrest Survivors," S. Laureys, ed., *Progress in Brain Research* 150 (2005): 356–57.

Karl Jansen, *Ketamine: Dreams and Realities* (Multidisciplinary Association for Psychedelic Studies, 2004): 134–35, 137–164.

B. B. Collier, "Ketamine and the Conscious Mind," *Anaesthesia* 27 (1972): 120–34.

A. Bianchi, "Comments on 'The Ketamine Model of the Near-Death Experience: A Central Role for the N-Methyl-D-Aspartate Receptor,'" *Journal of Near-Death Studies* 16, no. 1 (1997): 71–78.

D. K. Kim, "Ketamine Associated Psychedelic Effects and Dependence," *Singapore Medical Journal* 44, no. 1 (2003): 31–34.

R. E. Johnstone, "A Ketamine Trip," *Anesthesiology* 39 (1973): 460–61.

Rick Strassman, "Endogenous Ketamine-Like Compounds and the NDE," *Journal of Near-Death Studies* 16 (1997): 27–42.

Gerald Woerlee, *The Unholy Legacy of Abraham* (booklocker.com, 2008): 136–138, 282–289. Accessed October 26, 2010, www.unholylegacy.woerlee.org/images/unholy-legacy.pdf

Susan Blackmore, *Dying to Live*, ebook (Prometheus Books, 1993): Loc. 87–89, 616–57, 736–38, 1217–20, 1309–57, 2536–80.

Susan Blackmore, "Experiences of Anoxia: Do Reflex Anoxic Seizures Resemble NDEs?" *Journal of Near-Death Studies* 17 (1998): 111–120.

Bruce Greyson et al., "Explanatory Models for Near Death Experiences," *Handbook of Near Death Experiences* (Praeger, 2009): 219–20.

Sam Parnia, *What Happens When We Die?* (Hay House, 2006): 21. *Author's note*: I interviewed Parnia in February 2009.

Alex Tsakiris, "EEG Expert Can't Explain Near Death Experience Data," accessed October 26, 2010, http://www.skeptiko.com/eeg-expert-on-near-death-experience/

Janice Miner Holden, "Veridical Perceptions in Near-Death Experiences," and "Explanatory Models for Near Death Experiences," *Handbook of Near Death Experiences* (Praeger, 2009): 193–203, 30–231. *Author's note:* In this latter reference, Sabom constructed a particularly interesting experiment, in which experienced cardiac arrest patients were asked to describe their resuscitations and failed, inexperienced subjects who underwent NDEs proved accurate.

A quick quote from French:

"Challenges facing those proposing purely organic theories include not only producing direct evidence in support of their accounts, but also satisfactorily accounting for those NDEs that are known to occur in the complete absence of physical threat, such as those that occur when individuals are not actually close to death but only think they are."

Further, for a strong skeptical take, I highly recommend: Keith Augistine, "Hallucinatory Near-Death Experiences" (2003/Updated 2008) accessed

October 21, 2010, at http://www.infidels.org/library/modern/keith_augustine/HNDEs.html

Erik Davis, "Terence McKenna's Last Trip," *Wired* (May 2000), accessed October 21, 2010, http://www.wired.com/wired/archive/8.05/mckenna.html

Author's note: Susan Blackmore reveals herself rather fully in Barbara Bradley Hagerty's *Fingerprints of God,* saying "The idea of life after death is daft." So I guess Blackmore thinks she knows (Riverhead Books, 2009): 308.

Willoughby Britton et al., "Near-Death Experiences and the Temporal Lobe," *Psychological Science* 15, no. 4 (2004): 254–58.

Mario Beauregard and Denyse O'Leary, *The Spiritual Brain* (HarperOne, 2007): 68–76.

Willoughby Britton, Interview, January 2009.

Anahad O'Connor, "Following a Bright Light to a Calmer Tomorrow," *New York Times*, April 13, 2004.

Steven Kottler, "Extreme States," *Discover* (July 2005), accessed October 26, 2010, http://discovermagazine.com/2005/jul/extreme-states

David Paul Kuhn, "Both Parties Have Their Fanatics," *Real Clear Politics,* August 3, 2009, accessed October 21, 2010, http://www.realclearpolitics.com/articles/2009/08/03/each_party_has_its_fanatics_97748.html

Jim Geraghty, "55 Percent of Likely Voters Find 'Socialist' an Accurate Label of Obama," *National Review,* July 9, 2010, accessed, October 21, 2010, http://www.nationalreview.com/campaign-spot/230874/55-percent-likely-voters-find-socialist-accurate-label-obama

Nick Wing, "Poll: 35% of Republicans Want to Impeach Obama," *Huffington Post,* December 10, 2009, accessed October 21, 2010, http://www.huffingtonpost.com/2009/12/10/poll–35-of-republicans-wa_n_387093.html

Z. Klemenc-Ketis, "The Effect of Carbon Dioxide on Near-Death Experiences in Out-of-Hospital Cardiac Arrest Survivors," *Critical Care* 14, no. 2 (2010), accessed October 21, 2010, http://ccforum.com/content/14/2/R56

"Learn About Arterial Blood Gases," *Education for Nurses*, accessed October 21, 2010, http://www.the-abg-site.com/about.htm

John Lippman, "How Deep is TOO Deep?" *Divers Alert Network*, accessed October 21, 2010, at http://www.diversalertnetwork.org/medical/articles/article.asp?articleid=29

Jeff Wise, "Go Toward the Light," *Psychology Today*, April 16, 2010, Accessed online October 21, 2010, http://www.psychologytoday.com/blog/extreme-fear/201004/go-toward-the-light-the-science-near-death-experiences

Alex Tsakiris, "Do Science Journalists Get It Wrong," and Dr. Bruce Greyson, email to Alex Tsakaris, accessed October 21, 2010, http://www.skeptiko.com/near-death-experience-research-do-science-journalists-get-it-wrong/

Pim van Lommel, *Consciousness Beyond Life* (HarperOne, 2010): 116–17.

M. Morse, "Near Death Experiences: A Neurophysiological Explanatory Model," *Journal of Near-Death Studies* 8, (1989): 45–53.

Sam Parnia et al., "A Qualitative and Quantitative Study of the Incidence, Features and Aetiology of Near Death Experiences in Cardiac Arrest Survivors," *Resuscitation* 48, (2001): 149–56.

Michael Sabom, *Recollections of Death* (HarperCollins, 1982): 178.

Elisabeth Kübler-Ross, "Bio," accessed October 21, 2010, http://www.ekrfoundation.org/honorary-degrees

Facing Death: Elisabeth Kübler-Ross, directed by Stefan Haupt, DVD (First Run Features, 2007).

"Behavior: The Conversion of K," *Time*, November 12, 1979, accessed on October 21, 2010, at http://www.time.com/time/magazine/article/0,9171,946362–1,00.html

Karen G. Jackovich, "Sex, Visitors from the Grave, Psychic Healing: Kübler-Ross Is a Public Storm Center Again," *People,* October 29, 1979, accessed on October 21, 2010, http://www.people.com/people/archive/article/0,,20074920,00.html

Ron Rosenbaum, "Turn On, Tune In, Drop Dead," *Harpers* (September 1982). Also available in Rosenbaum's book, *The Secret Parts of Fortune* (Perennial, 2000): 253–67.

Elisabeth Kübler-Ross, *AIDS: The Ultimate Challenge* (Scribner, 1997).

CHAPTER 2: DO YOU SEE WHAT I SEE?

Brian Josephson, "Physics and the Nobel Prizes," Royal Mail special stamps booklet, October 2, 2001, accessed October 22, 2010, http://www.tcm.phy.cam.ac.uk/~bdj10/stamps/text.html

Robin McKie, "Royal Mail's Nobel Guru in Telepathy Row," *Observer*, September 30, 2001, accessed October 22, 2010, http://www.guardian.co.uk/uk/2001/sep/30/robinmckie.theobserver

Herodotus, *The History*, trans. David Grene (Univ. of Chicago Press, 1987): 53–54.

B. Josephson, "The Discovery of Tunnelling Supercurrents," *Review of Modern Physics* 46, no. 2 (1974): 251–54.

Edward Edelson, "Mammoth Magnets to Microchips from Superconductors," *Popular Science* 152 (May 1981): 73–79.

Robert McDermott et al., "Microtesla MRI with a Superconducting Quantum Interference Device," *Proceedings of the National Academy of Sciences* 101, no. 21 (May 25, 2004): 7857–61.

BBC Radio Interview, October 2, 2001, accessed on October 22, 2010, http://www.tcm.phy.cam.ac.uk/BIG/bdj10/audio/stamps.mp3

Danny Penman, "Is this Proof We're all Psychic," *Daily Mail*, January 28, 2008, accessed on October 22, 2010, http://www.dailymail.co.uk/news/article–510762/Could-proof-theory-ALL-psychic.html

podblack, *PodBlack Cat,* blog, "Dr. Richard Wiseman on Remote Viewing in the

Daily Mail—Clarification," September 28, 2009, http://podblack.com/2009/09/dr-richard-wiseman-on-remote-viewing-in-the-daily-mail-clarification/

Christopher French, Interviews, January, February 2010.

Parapsychological Association Convention, Thursday August 6, 2009–Sunday August 9, 2009. Material gathered at the conference is identified as such within the text.

Paul H. Smith, Interview, August 2010.

Dean Radin, *The Conscious Universe* (HarperOne, 1997): 13–24.

psi definition, in physics, taken from an online collection of academic dictionaries: http://en.academic.ru/dic.nsf/enwiki/238149

Paul D. Allison, "Experimental Parapsychology as a Rejected Science," *On the Margins of Science: The Social Construction of Rejected Knowledge. The Sociological Review Monograph,* 27 (1979): 271–91.

Author's note: My own survey yielded a far lower response rate than Allison's. His was 90 percent, mine was around 25 percent. I'll make no scientific claims for the worthiness of my study, but my own experience at the parapsychology conference so closely mirrored Allison's original findings, and my survey results were so similar to his, I feel extremely confident in the accuracy of what I've reported here. That said, readers should feel free to accept or reject my analysis purely on the basis of their own preconceived biases. (That last line is a joke.)

James H. Lee, "Remote Viewing as Applied to Futures Studies," *Technological Forecasting and Social Change* 75, no. 1 (2008): 142.

Ken Kress, "Parapsychology in Intelligence," *Journal of Scientific Explanation* 13, no. 1, (1999): 68–85.

Author's note: Kress, a CIA analyst, evaluates the evidence for psi and takes a mixed view of Price. Though most skeptics might like to focus on the negative, the most important line is buried toward the end of Kress's report: "There are observations, such as the original magnetic experiments at Stanford University, the OSI remote viewing, the OTS-coderoom experiments, and others done for the Department of Defense, that defy explanation. Coincidence is not likely, and fraud has not been discovered. The implication of these data cannot be determined until the assessment is done. If the above is true, how is it that the phenomenon remains controversial and receives so little official government support? . . . This state of affairs occurs because of the elementary understanding of parapsychology and because of the peculiarities of the intelligence and military organizations which have attempted the assessments. There is no fundamental understanding of the mechanisms of paranormal functioning, and the reproducibility remains poor."

Paul H. Smith, *Reading the Enemy's Mind* (Forge, 2005): 128–29.

Ray Hyman, "Evaluation of a Program on Anomalous Mental Phenomena," *Journal of Scientific Exploration* 10, no. 1 (1996): 31–58.

Dean Radin, *Entangled Minds* (Pocket Books, 2006): 120–21.

What follows is a short selection of papers on psi research I found most compelling during my research:

Joel B. Greenhouse, "Comment: Parapsychology—On the Margins of Science," *Statistical Science* 6, no. 4 (1991), accessed October 26, 2010, http://www.stat.ucdavis.edu/~utts/91rmp-c4.html

Jessica Utts and Brian Josephson, "The Paranormal: The Evidence and Its Implications for Consciousness," 1996, accessed October 26, 2010, http://www.tcm.phy.cam.ac.uk/~bdj10/psi/tucson.html

George P. Hansen, "The Elusive Agenda: Dissuading as Debunking in Ray Hyman's *The Elusive Quarry*," *Journal of the American Society for Psychical Research* 85 (April 1991): 193–203, accessed October 26, 2010, http://www.tricksterbook.com/ArticlesOnline/HymanReview.htm

Jessica Utts, "Replication and Meta-Analysis in Parapsychology," *Statistical Science* 6, no. 4 (1991): 363–403.

Bernard Carr, "Rational Perspective on the Paranormal," Conference Report, *Journal of Scientific Exploration,* 16, no. 4 (2002): 635–50.

Richard Wiseman and Marilyn Schlitz, "Experimenter Effects and the Remote Detection of Staring," *Journal of Parapsychology* 61, no. 3 (1998): 197–208.

Daryl Bem and Charles Honorton, "Does Psi Exist? Replicable Evidence for an Anomalous Process of Information Transfer," *Psychological Bulletin* 115, no. 1 (1994): 4–18.

J. Milton and R. Wiseman, "Does Psi Exist? Lack of Replication of an Anomalous Process of Information Transfer," *Psychological Bulletin* 125, no. 4 (1999): 387–91.

L. Storm and S. Ertel, "Does Psi Exist? Comments on Milton and Wiseman's Meta-Analysis of Ganzfeld Research," *Psychological Bulletin* 127, no. 3 (2001): 424–33.

Daryl Bem et al., "Updating the Ganzfeld Database: A Victim of Its Own Success?" *Journal of Parapsychology* 65 (2001): 207–18.

L. Storm et al.,"Meta-Analysis of Free-Response Studies, 1992–2008: Assessing the Noise Reduction Model in Parapsychology," *Psychological Bulletin* 136, no. 4 (July 2010): 471–85.

Guy Lyon Playfair, "Twin Telepathy," *Fortean Times Paranormal Handbook* (2009): 70–75.

See Radin, *Entangled Minds,* 136–41, and papers Radin collected and analyzed at http://deanradin.blogspot.com/ for brain correlation experiments, listed here.

C. Tart, "Possible Physiological Correlates of psi Cognition," *International Journal of Parapsychology* 5 (1963): 375–86.

T. D. Duane et al., "Extrasensory Electroencephalographic Induction Between Identical Twins," *Science* 150 (1965): 367.

D. H. Lloyd et al., "Objective Events in the Brain Correlating with Psychic Phenomena," *New Horizons* 1 (1973): 69–75.

C. S. Rebert et al., "EEG Spectrum Analysis Techniques Applied to the Problem of psi Phenomena," *Behavioral Neuropsychiatry* 6 (1974): 18–24.

Russell Targ and Hal Puthoff, "Information Transmission Under Conditions of Sensory Shielding," *Nature* 252 (1974): 602–7.

B. Millar et al., "An Attempted Validation of the 'Lloyd Effect,'" *Multidimensional Mind: Remote Viewing in Hyperspace* (Scarecrow Press, 1999): 25–27.

E. F. Kelly, "EEG Changes Correlated with a Remote Stroboscopic Stimulus: A Preliminary Study," *Research in Parapsychology* (Scarecrow Press, 1975): 58–63.

K. Hearne, "Visually Evoked Responses and ESP," *Journal of the Society for Psychical Research* 49 (1977): 648–57.

Z. Vassy, "Method for Measuring the Probability of 1 Bit Extrasensory Information Transfer Between Living Organisms," *Journal of Parapsychology* 42 (1978): 158–60.

K. Hearne, "Visually Evoked Responses and ESP: Failure to Replicate Previous Findings," *Journal of the Society for Psychical Research* 51 (1981): 145–47.

D. W. Orme-Johnson et al., "Intersubject EEG Coherence: Is Consciousness a Field?" *International Journal of Neuroscience* 16 (1982): 203–9.

J. Grinberg-Zylberbaum et al., "Patterns of Interhemispheric Correlation During Human Communication," *International Journal of Neuroscience* 36 (1987): 41–53.

J. Grinberg-Zylberbaum et al., "The Einstein-Podolsky-Rosen Paradox in the Brain: The Transferred Potential," *Physics Essays* 7 (1994): 422–28.

H. Sugano et al., "A New Approach to the Study of Subtle Energies," *Subtle Energies* 5, no. 2 (1994): 143–65.

C. N. Shealy et al., "EEG Alterations During Absent 'Healing,'" *Subtle Energies* 11, no. 3 (2000): 241–48.

A. Sabell et al., "Inter-Subject EEG Correlations at a Distance—The Transferred Potential," *Proceedings of the 44th Annual Convention of the Parapsychological Association* (2001): 419–22.

H. Walach et al., "Transferred Potentials—Fact or Artifact? Results of a Pilot Study. In Bridging Worlds and Filling Gaps in the Science of Healing," *Samueli Institute for Information Biology* (2001): 303–25.

E. C. May et al., "EEG Correlates to Remote Light Flashes Under Conditions of Sensory Shielding," *Mind at Large: IEEE Symposia on the Nature of Extrasensory Perception* (Hampton Roads, 2002).

S. Kalitzin et al., "Comments on 'Correlations Between Brain Electrical Activities of Two Spatially Separated Human Subjects,'" *Neuroscience Letters* 350, no. 3 (October 30, 2003): 193–94.

L. Standish et al., "Evidence of Correlated Functional MRI Signals Between Distant Human Brains," *Alternative Therapies in Health and Medicine* 9 (2003): 122–28.

J. Wackermann et al., "Correlations Between Brain Electrical Activities of Two Spatially Separated Human Subjects. Reply to the Commentary by S. Kalitzin and P. Suffczynski," *Neuroscience Letters* 350, no. 3 (2003): 193–94.

J. Wackerman, et al., "Correlations Between Brain Electrical Activities of Two Spatially Separated Human Subjects," *Neuroscience Letters* 336 no. 1 (2003): 60–64.

U. Hasson, "Intersubject Synchronization of Cortical Activity During Natural Vision," *Science* 303 (2004): 1634–40.

M. Kittenis et al., "Distant Psychophysiological Interaction Effects Between Related and Unrelated Participants," *Proceedings of the Parapsychological Association Convention* (Vienna, Austria, August 5–8, 2004): 67–76.

Dean Radin, "Event-Related Electroencephalographic Correlations Between Isolated Human Subjects," *Journal* of *Alternative and Complementary Medicine* 10 no. 2 (2004): 315–23.

L. Standish et al. "Electroencephalographic Evidence of Correlated Event-Related Signals Between the Brains of Spatially and Sensory Isolated Human Subjects," *Journal of Alternative and Complementary Medicine* 10, no. 2 (2004): 307–14.

S. Schmidt, "Distant Intentionality and the Feeling of Being Stared At: Two Meta-Analyses," *British Journal of Psychology* 95 (2004): 235–47.

J. Wackerman, "Dyadic Correlations Between Brain Functional States: Present Facts and Future Perspectives," *Mind and Matter* 2, no.1 (2004): 105–22.

J. Achterberg et al., "Evidence for Correlations Between Distant Intentionality and Brain Function in Recipients: a Functional Magnetic Resonance Imaging Analysis," *Journal of Alternative and Complementary Medicine* 11, no. 6 (2005): 965–71.

T. L. Richards et al., "Replicable Functional Magnetic Resonance Imaging Evidence of Correlated Brain Signals Between Physically and Sensory Isolated Subjects," *Journal of Alternative and Complementary Medicine* 11, no. 6 (2005): 955–63.

S. T. Moulton et al., "Using Neuro-Imaging to Resolve the psi Debate," *Journal of Cognitive Neuroscience* 20, no. 1 (2008): 182–92.

Michael Persinger et al., "Enhanced Power Within a Predicted Narrow Band of Theta Activity During Stimulation of Another by Circumcerebral Weak Magnetic Fields After Weekly Spatial Proximity: Evidence for Macroscopic Quantum Entanglement?" *NeuroQuantology* 6, no. 1 (2008): 7–21.

B. T. Dotta, "Evidence of Macroscopic Quantum Entanglement During Double Quantitative Electroencephalographic Measurements of Friends vs Strangers," *NeuroQuantology* 7, no. 4 (2009): 548–51.

Sam Harris, *The End of Faith* (Norton, 2004): 41.

Michael Persinger, Interview, June 2009.

Michael Persinger, "The Harribance Effect as Pervasive Out-of-Body Experiences: NeuroQuantal Evidence with More Precise Measurements," 8, no. 4 (2010): 444–465.

Jessica Utts, Interview, August 2009. *Author's note:* Utts believes the PK tests reflect precognition—not the mental manipulation of matter, as PK enthusiasts claim.

Jessica Utts, *Seeing Through Statistics* (Brooks/Cole, 1999).

Marcello Truzzi, "On Some Unfair Practices Towards Claims of the Paranormal," *Oxymoron: Annual Thematic Anthology of the Arts and Sciences, Vol. 2: The Fringe* (Oxymoron Media, 1998). Accessed October 26, 2010, http://www.skepticalinvestigations.org/Anomali/practices.html

David Hume, "Of Miracles," *An Enquiry Concerning Human Understanding* (Filiquarian Publishing, 2007): 104.

Ray Hyman, "The Evidence for Psychic Functioning: Claims vs. Reality," *Skeptical Inquirer* 20, no. 2 (March/April, 1996), accessed October 26, 2010, http://www.csicop.org/si/show/evidence_for_psychic_functioning_claims_vs._reality/

Trevor Pinch, Interview, September 2009.

Trevor J. Pinch, "Normal Explanations of the Paranormal," *Social Studies of Science* 9 (1979): 329–48.

Trevor Pinch, "Private Science and Public Knowledge: The Committee for the Scientific Investigation of the Claims of the Paranormal and Its Use of the Literature," *Social Studies of Science* 14, (1984): 521.

H. M. Collins and Trevor Pinch, *Frames of Meaning: The Social Construction of Extraordinary Science* (Routledge, 1982).

Elizabeth Mayer, *Extraordinary Knowing* (Bantam, 2007): 69–70, 93.

Marie-Catherine Mousseau, "Parapsychology: Science or Pseudo-Science?" *Journal of Scientific Exploration* 17, no. 2 (2003): 271–82.

Chris Carter, *Parapsychology and the Skeptics* (Paja Books, 2007): 8–15, 73–82. *Author's note*: Carter's book on the battle between skeptics and psi proponents is, for my money, one of the best books ever written about the paranormal. My citation here is for the Rawlins and Wiseman material, but Carter's entire book is worth reading and was an incredible resource for me in the research for this chapter. A further chapter of Carter's book, available online, is cited later.

George P. Hansen, "CSICOP and the Skeptics: An Overview," *Journal of the American Society for Psychical Research* 86, no. 1 (January 1992): 19–63.

Guy Lyon Playfair, "Has CSICOP Lost the Thirty Years' War?" *Skeptical Investigations*, accessed October 26, 2010, http://www.skepticalinvestigations.org/New/Observeskeptics/CSICOP/30yearswar1.html

Paul Kurtz et al., "Objections to Astrology," *Humanist* (September/October 1975): 4–6.

Dennis Rawlins, "Starbaby," *Fate* 34, no. 10 (October 1981): 67–98. Accessed October 26, 2010, http://cura.free.fr/xv/14starbb.html

Michel and Francoise Gauquelin, "The Truth About the Mars Effect on Sports Champions," *Humanist* 36, no. 4 (July/August 1976): 44–55.

G. O. Abell et al., "A Test of the Gauquelin Mars Effect," *Humanist* 36, no. 5 (September/October 1976): 40.

Michel and Francoise Gauquelin, "The Zelen Test of the Mars Effect," *Humanist* 37, no. 6 (November/December 1977): 30–35.

Richard Kammann, "The True Disbelievers: Mars Effect Drives Skeptics to Irrationality," *Zetetic Scholar* 10 (1982): 50–65.

Patrick Curry, "Research on the Mars Effect," *Zetetic Scholar* 9 (February/March 1982): 34–52.

R. Targ and H. Puthoff, "Information Transmission Under Conditions of Sensory Shielding," *Nature* 251, no. 18 (October 1974): 602–7.

James Randi, *Flim-Flam!* (Prometheus Books, 1982): 133 (lying claim), 143–45 (Geller tests, Pressman controversy).

Guy Lyon Playfair, "The Witch Hunters," *Geller Effect*, 1988, Accessed October 26, 2010, http://www.urigeller.com/geller-effect/tge14.htm

Jonathan Margolis, *Uri Geller: Magician or Mystic?* (Orion, 1999).

Geller has posted the entire book for free at http://www.uri-geller.com/books/magician-or-mystic/index.htm, but the relevant chapter was accessed October 26, 2010, http://www.uri-geller.com/books/magician-or-mystic/chapter11.htm.

Paul H. Smith, *Reading the Enemy's Mind* (Forge, 2005): 66–67.

Susan Blackmore, *In Search of the Light: Adventures of a Parapsychologist* (Prometheus Books, 1986): 163.

Susan Blackmore, "The Elusive Open Mind: Ten Years of Negative Research in Parapsychology," *Skeptical Inquirer*, 11, (1987): 244–55. *Author's note:* Some might complain that I subject Blackmore to a particularly critical examination. In that context, I congratulate her on the following article she published in 2010, in which she reveals that she no longer considers religion a "virus of the mind." In it, she cites some new research that has come to her attention demonstrating that religious people seem happier, perhaps even healthier, and have more children than "secularists." I'm not sure why she is shocked by this new data when plenty of old data contained similar findings, but I congratulate her on being willing to revise her opinion on such a hot button issue. Susan Blackmore, "Why I No Longer Believe Religion is a Virus of the Mind," *Guardian*, September 16, 2010, accessed October 30, 2010, http://www.guardian.co.uk/commentisfree/belief/2010/sep/16/why-no-longer-believe-religion-virus-mind/print.

Rick E. Berger, "A Critical Examination of the Blackmore Psi Experiments," *Journal of the American Society for Psychical Research* 83 (1989): 123–44.

Chris Carter, "The Research of the Skeptics," *Skeptical Investigations*. Also *Parapsychology and the Skeptics*, 70–71, accessed October 26, 2010, http://www.skepticalinvestigations.org/Anomali/skeptic_research.html

Susan Blackmore, "A CRITICAL RESPONSE to Rick Berger," *Journal of the American Society for Psychical Research* 83 (1989): 145–54.

Sheila Jones, "One Hot-Button Issue Can Rile a Roomful of Skeptics," *Globe and Mail*, July 10, 2010, p. F-4.

Susan Blackmore, "Which Skeptical Position?" *Skeptical Inquirer* 19, no. 3 (May/June 1995): 26.

Alex Tsakiris, "Rupert Sheldrake and Richard Wiseman Clash Over Parapsychology Experiments," *Skeptiko*, March 8, 2010, http://www.skeptiko.com/rupert-sheldrake-and-richard-wiseman-clash/. Accessed October 26, 2010.

Also see, Carter, "Research."

Kendrick Frazier, "It's CSI Now, Not CSICOP," December 4, 2006. Accessed October 26, 2010. http://www.csicop.org/specialarticles/show/its_csi_now_not_csicop/

Brian Josephson, "Scientists' Unethical Use of Media for Propaganda Purposes," Cavendish Laboratory, Cambridge University, Fall 2004, accessed October 26, 2010, http://www.tcm.phy.cam.ac.uk/~bdj10/propaganda/

Guy Lyon Playfair, "The Girl with X-Ray Eyes," *Skeptical Investigations*, accessed October 26, 2010, http://skepticalinvestigations.org/Demkinafile/X-Ray.html

Rupert Sheldrake, "James Randi—A Conjurer Attempts to Debunk Research on Psychic Animals," accessed October 26, 2010, http://www.sheldrake.org/D&C/controversies/randi.html

Brandon K. Thorp, "The Sheldrake Kerfuffle," *Swift*, blog of the James Randi Educational foundation, accessed October 26, 2010, http://www.randi.org/site/index.php/swift-blog/795-the-sheldrake-kerfluffle.html

James Randi, "Nessie Innocent, Sheldrake Lecture, Hand Acupressure, More Santa Stuff, and a Good Christian's Dilemma," *Swift*, January 17, 2003, accessed October 26, 2010, http://www.randi.org/jr/011703.html

Author's note: The following selections, with some notes from me, contain some of the back-and-forth between Schwartz and Randi.

James Randi, "How Long Do We Wait," *Swift*, April 8, 2005. *Author's note*: Here, Randi lists Krippner, in a March 27, 2001 (author's emphasis) letter to Schwartz's university, as having "already agreed" to analyze Schwartz's data, Accessed October 26, 2010, http://www.randi.org/jr/040805how.html

Gary Schwartz, "A Reply to Randi," *Daily Grail*, April 15, 2005, accessed October 26, 2010, http://www.dailygrail.com/Guest-Articles/2005/4/Reply-Randi-Dr-Gary-Schwartz Two months after the letter listing Krippner, Randi says Krippner *didn't* agree to participate and this was only his proposed panel.

James Randi, " . . . More Schwartz!" *Swift*, May 18, 2001. Accessed January 11, 2011. http://www.randi.org/site/index.php/swift-blog/swift-archives.html

Michael Schmicker, *Best Evidence* (Writers Club Press, 2002): 287–88.

Ridolfo H. Baxter et al., "Social Influences on Paranormal Belief," *Current Research in Social Psychology* 15, no. 3 (2010) 33-41. *Author's note:* Also includes polling data.

Frank Newport et al., "Americans' Belief in Psychic and Paranormal Phenomena Is up Over Last Decade," Gallup News Service, June 8, 2001.

David W. Moore, "Three in Four Americans Believe in Paranormal," Gallup News Service, June 16, 2005.

Daryl J. Bem, et al., "Does Psi Exist? Replicable Evidence for an Anomalous Process of Information Transfer," *Psychological Bulletin* 115, no. , (1994): 4–18.

Dean Radin, *The Conscious Universe* (HarperOne, 1997): 55–56.

Doreen Molloy, Personal reading, January 2010.

Windbridge Certified Research Mediums, accessed October 26, 2010, http://www.windbridge.org/mediums.htm

Julie Beischel et al., "Anomalous Information Reception by Research Mediums Demonstrated Using a Novel Triple-Blind Protocol," *EXPLORE: The Journal of Science and Healing* 3, no. 1 (2007): 23–27.

M. Mumford et al., "An Evaluation of Remote Viewing: Research and Applications," *American Institutes for Research*, September 29, 1995.

Patrick Huyghe, "Closing the Dream Factory: Sony Proves That Psychic Powers Are Real," *Fortean Times*, October 1998.

CHAPTER 3: OUT OF THEIR HEADS? OFF WITH THEIR HEADS!

Roger Penrose, *The Emperor's New Mind*, (1989, Oxford Univ. Press): 4 (initial quote), 75–84 (Turing machines, beginning of Gödel's theorem argument), 138–46 (full Godel's theorem argument), 516–17 (QM in the brain, photon experiment), 521 (need for a new physics to explain consciousness), 523–581 (the physics of the mind—a brilliant summation of his position).

Stuart Hameroff, Interview, March 12–14 2009.

Beyond Belief: Science, Reason, Religion & Survival, conference, November 5, 2006; Hameroff's presentation can be seen at http://thesciencenetwork.org/programs/beyond-belief-science-religion-reason-and-survival/session–4–1

Stuart H. Hameroff, "The Entwined Mysteries of Anesthesia and Consciousness: Is There a Common Underlying Mechanism?" *Anesthesiology* 105 (2006): 400–412.

G. A. Mashour, "Integrating the Science of Consciousness and Anesthesia," *Anesthesia and Analgesia* 103 (2006): 975–82.

G. A. Mashour, "The Cognitive Binding Problem: From Kant to Quantum Neurodynamics," *NeuroQuantology* 1 (2004): 29–38.

Stuart Hameroff, *Ultimate Computing* (Elsevier Science, 1987) can be downloaded for free: http://www.quantumconsciousness.org/ultimatecomputing.html

Jordan Goodman and Vivien Walsh, The Story of Taxol: Nature and Politics in the pursuit of an anti-cancer drug, (2001): 90.

P. Drabik et Al., "Microtubule Stability Studied by Three-Dimensional Molecular Theory of Solvation," *Biophysical Journal* 92, 2 (2007): 394–403.

J. Fabera et al., "Information Processing in Brain Microtubules," *BioSystems* 83 (2006): 1–9.

Stuart Hameroff and Richard C. Watt, "Information Processing in Microtubules," *Journal of Theoretical Biology* 98 (1982): 549–61. *Author's note:* This is where references to *The Emperor's New Mind* are most prevalent—see earlier note for Penrose, *Emperor's*.

David H. Freedman, "Quantum Consciousness," *Discover* (June 1994): 89–98.

Thanks to physicists Jack Tuszynski and Jim Trolinger for vetting my write-up of the (very) basics of quantum physics.

H. Schmidt et al., "Channeling Evidence for a PK Effect to Independent Observers," *Journal of Parapsychology* 50 (March 1986): 1–15. *Author's note:* Schmidt claims his tests suggest the role of a conscious observer—a human being—are necessary to collapse the quantum wave function.

A. Aspect, "Experiments on Einstein-Podolsky-Rosen-type Correlations with Pairs of Visible Photons," *Quantum Concepts in Space and Time* (1986).

Vincent Jacques, E. Wu, Frdric Grosshans, Franois Treussart, Philippe Grangier, Alain Aspect, Jean-Franois Roch, "Experimental Realization of Wheeler's Delayed-Choice Experiment," *Science* 315 (2007): 5814.

"Copenhagen Interpretation of Quantum Mechanics," *Stanford Encyclopedia of Philosophy*. *Author's note:* I found this relatively brief account particularly clear. It can be found at http://plato.stanford.edu/entries/qm-copenhagen/.

"Many-Worlds Interpretation of Quantum Mechanics," *Stanford Encyclopedia of Philosophy*, http://plato.stanford.edu/entries/qm-manyworlds/#6.4.

Peter Byrne, "The Many Worlds of Hugh Everett," *Scientific American* (December 2007), accessed October 26, 2010, http://www.scientificamerican.com/article.cfm?id=hugh-everett-biography

Max Tegmark, "The Interpretation of Quantum Mechanics, Many Worlds or Many Words," *Fortschritte der Physik* 46 (1998): 855–62.

Tim Radford, "David Deutsch's Multi-Verse Carries Us Beyond the Realms of Imagination," *Guardian*, June 11, 2010.

R. Courtland, "Infinite Doppelgängers May Explain Quantum Probabilities," *New Scientist*, August 26, 2010.

Jim Elvidge, *The Universe—Solved!* (AT Press, 2008): 35–36.

Frank J. Tipler, *The Physics of Immortality* (Anchor Books, 1995): 170–71.

Author's note: The references located immediately above contain further information on the Raub poll, but to ascertain the depth of Hawking's commitment to a many-worlds or multiverse theory (I write *a* because there are variations), one need look no further than his recent book, *The Grand Design* (Bantam, 2010).

Richard Feynman, *QED: The Strange Theory of Light and Matter* (Princeton Univ. Press, 1988): 5.

Einstein, in a letter to physicist Max Born, 1924, in F. Shapiro and J. Epstein, *Yale Book of Quotations* (Yale Univ. Press, 2006): 228.

Max Born and Albert Einstein, *The Born-Einstein Letters 1916–1955* (Macmillan, 2004): 80.

Victor Stenger, "Quantum Quackery," *Skeptical Inquirer* 21, no. 1 (January/February 1997), accessed October 30, 2010, http://www.csicop.org/si/show/quantum_quackery

Stuart Hameroff, "Naughty Quantum Robot," http://www.quantumconsciousness.org/interviews/objectmonkey.html

Stuart Hameroff and Roger Penrose, "Orchestrated Reduction of Quantum Coherence in Brain Microtubules: A Model for Consciousness?" *Toward a Science of Consciousness—The First Tucson Discussions and Debates* (MIT Press, 1996): 507–40.

Max Tegmark, "Importance of Quantum Coherence in Brain Processes," *Physical Review E* 61 (2000): 4194–206.

S. Hagan et al., "Quantum Computation in Brain Microtubules: Decoherence and Biological Feasibility," *Physical Review E* 65 (2002): DOI: 10.1103/PhysRevE.65.061901.

Rick Grush and Patricia Churchland, "Gaps in Penrose's Toilings," *Journal of Consciousness Studies* 2, no. 1 (1995): 10–29.

Patricia Churchland, "Brainshy: Nonneural Theories of Conscious Experience," *Toward a Science of Consciousness II: The Second Tucson Discussions and Debates* (MIT Press, 1998): 109–24.

Roger Penrose and Stuart Hameroff, "Gaps, What Gaps? Reply to Grush and Churchland," *Journal of Consciousness Studies* 2, no. 2 (1995): 99–112.

Stuart Hameroff, "More Neural Than Thou (A Reply to Patricia Churchland)," *Toward a Science of Consciousness II: The Second Tucson Discussions and Debates* (MIT Press, 1998): 197–213.

Author's note: I am well aware that there are more papers than the ones mentioned that discuss the Penrose-Hameroff model for consciousness. But reading over the back-and-forth between Hameroff and his opponents, it seems the matter remains open. My own take is that when Penrose himself turns back to this theory and reviews all that has happened since he put it forward, some real action will commence.

Brian Greene, *Fabric of the Cosmos* (Vintage, 2004): 351.

Lisa Randall and Art Bell, Interview, *Coast to Coast A.M.*, February 25, 2006.

Mark Buchanan, "Do Birds See with Quantum Eyes?" *New Scientist* (May, 3, 2008), accessed October 30, 2010, http://www.newscientist.com/article/mg19826544.000-do-birds-see-with-quantum-eyes.html

"Quantum Biology Has Come In from the Cold," editorial, *New Scientist* (February 2010), accessed October 30, 2010, http://www.newscientist.com/article/mg20527462.500-quantum-biology-has-come-in-from-the-cold.html

Gregory S. Engel et al., "Evidence for Wavelike Energy Transfer Through Quantum Coherence in Photosynthetic Systems," *Nature* 446 (April 2007): 782–86.

A. Zeilinger, "Quantum Interference Experiments with Large Molecules," *American Journal of Physics* 71, no. 4, (2003): 319–25.

M. Arndt et al., "Quantum Physics Meets Biology," *HFSP Journal*, 3, no. 6 (December 2009): 386–400.

J. C. Brookes et al., "Could Humans Recognize Odor by Phonon Assisted Tunneling?" *Physical Review Letters* (2007): 98.

Geoff Brumfiel, "Scientists Supersize Quantum Mechanics: Largest Ever Object Put into Quantum State," *Nature* (March 2010), accessed October 30, 2010, http://www.nature.com/news/2010/100317/full/news.2010.130.html

J. R. Minkel, "Is Sense of Smell Powered by Quantum Vibrations? Controversial theory Gets Green Light from Physicists," *Scientific American* (December 15, 2006), accessed October 30, 2010, http://www.scientificamerican.com/article.cfm?id=is-sense-of-smell-powered

Elio Conte et al., "On the Existence of Quantum Wave Function and Quantum Interference Effects in Mental States: An Experimental Confirmation during Perception and Cognition in Humans," *NeuroQuantology* 7, no 2. (2009), accessed October 26, 2010, arxiv.org/pdf/0807.4547.

E. Conte, "Mental States Follow Quantum Mechanics During Perception and Cognition of Ambiguous Figures," *Open Systems and Information Dynamics* 16, no. 1 (2009): 85–100.

F. Beck, "Quantum Aspects of Brain Activity and the Role of Consciousness," *Proceedings of the National Academy of Science* 89 (1992):11357–361.

F. T. Arechhi, "Chaotic Neuron Dynamics, Synchronization, and Feature Binding: Quantum Aspects," *Mind and Matter* 1, no. 1 (2004): 15–43.

Mario Livio, "The Golden Ratio and Aesthetics," *Plus* 22 (November 2002), October 26, 2010. http://plus.maths.org/issue22/features/golden/

R. Coldea, "Quantum Criticality in an Ising Chain: Experimental Evidence for Emergent E8 Symmetry," *Science* 327, no. 5962 (January 8, 2010): 177–80.

"Golden Ratio Discovered in Quantum World: Hidden Symmetry Observed for the First Time in Solid State Matter," *Science Daily*, January 7, 2010, accessed October 26, 2010, http://www.sciencedaily.com/releases/2010/01/100107143909.htm

D. A. Baylor, "Responses of Retinal Rods to Single Photons," *Journal of Physiology* 288, (1979): 613–34.

S. Hecht, "Energy, Quanta and Vision," *Journal of General Physiology* 25 (1942): 891–940.

J. Roebke, "The Reality Tests," *Seed* (June 2008): 50–59.

"Spooky Action and Beyond," *Sign and Sight*, February 16, 2006, accessed October 30, 2010, http://www.signandsight.com/features/614.html

"Talking Physics with the Dalai Lama," *Physics World*, August 7, 1998, accessed October 30, 2010, http://physicsworld.com/cws/article/news/3186

A. Zajonc, ed. *The New Physics and Cosmology, Dialogues with the Dalai Lama* (Oxford Univ. Press, 2004): 11–30.

Thomas Campbell, *My Big Toe* (Lightning Strike Books, 2003).

B. Rosenblum and F. Kuttner, *Quantum Enigma* (Oxford Univ. Press, 2006): 6–7, 156–57, 183–208.

Hans Peter Dürr, "Matter Is Not Made Out of Matter," *Endogenous Development and Biocultural Diversity* (Compas, 2007): 45–55. *Author's note*: The depth of disagreement among *physicists*—never mind mystics—is perhaps best realized by perusing this talk, given by German physicist Hans Peter Dürr, a student of Werner Heisenbrg, and former executive director at the prestigious Max Planck Institute. Dürr, like Zeilinger, believes quantum mechanics might ultimately force us to a completely new understanding of reality; to him, that answer lies beyond materialism.

Roger Penrose, Email, December 2009.

"Professor Sir Roger Penrose," British Humanist Association, accessed October 26, 2010, http://www.humanism.org.uk/about/people/distinguished-supporters/roger-penrose-frs

David Chalmers, "Facing Up to the Problem of Consciousness," *Toward a Science of Consciousness* (MIT Press, 1996): 5–28.

D. D. Hoffman, "Conscious Realism and the Mind-Body Problem," *Mind & Matter* 6, no. 1 (2008): 87–121.

David Chalmers, Interview, November 2009.

A. Marshall Stoneham, Email, March 2009.

Arnaud Delorme, Interview, March 2010.

Daniel Dennett, *Consciousness Explained* (Little Brown, 1991).

Susan Blackmore, *Conversations on Consciousness* (Oxford Univ. Press, 2005): 79–82, 116–17.

Nancy Woolf, Interview, February 2010.

Jack Tuszynski, Interview, February 2010.

CHAPTER 4: BLAZING SADDLES

Jewel, "Stephenville, TX," *Goodbye Alice in Wonderland*, CD (Atlantic, 2009).

Lee Roy Gaitan, Interview, September 2009.

Tarleton State Univ., accessed October 27, 2010, http://www.tarleton.edu/about/Stephenville.html

Angelia Joiner, "Japan Interested in U.F.O. Sighting," *Stephenville Empire-Tribune*, January 25, 2008. *Author's note*: The *Empire-Tribune*'s articles are available behind a pay wall at http://www.empiretribune.com/.

Following is a selection of videos that attest to the media attention Stephenville received, accessed October 27, 2010.

http://www.youtube.com/watch?v=dqdX-iwk5Mc

http://www.youtube.com/watch?v=kPBmeX6pBgI&feature=related

http://www.youtube.com/watch?v=gBhjmf2JSuA&feature=related

Denise Gellene, "How UFOs Took Over a Town," *Los Angeles Times*, June 14, 2008, accessed October 30, 2010, http://articles.latimes.com/2008/jun/14/science/sci-ufo14

Sara Vanden Berge, Interview, April 2009.

Steve Allen, Interviews, September 2009, August 2010.

Weather data (hourly) obtained from Weather Warehouse, an online historical weather site, accessed October 30, 2010, http://weather-warehouse.com/

Angelia Joiner, Interview, August, September, and November 2009.

Glen Schulze and Robert Powell, "Special Research Report, Stephenville, Texas," MUFON, (July 2008): 6–7, 33–34.

Angelia Joiner, "Possible U.F.O Sighting," *Stephenville Empire-Tribune*, January 10, 2008.

Rick Sorrells, Interview, September 2009.

Angelia Joiner, "Dozens in Texas Town Report Seeing U.F.O.," *Stephenville Empire-Tribune*, January 14, 2008.

Angelia Joiner, "Three Erath County Lawmen Observe 'One Big Craft,'" *The Stephenville Lights*, March 30, 2008, accessed November 1 2010, http://stephenvillelights.com/slnews_article005.html

Angelia Joiner, "Stephenville UFO Is Viewed by Former Protector of Texas Governors," *The Stephenville Lights*, February 28, 2008, http://stephenvillelights.com/slnews_article004.html

Angelia Joiner, "All Eyes on the Skies," *Stephenville Empire-Tribune*, January 13, 2008.

Alejandro Rojas, Former Director of Public Education, MUFON, Interview, July 2009.

Robert Powell, Director of Research, MUFON, Interviews, July, August, and October 2009.

Jane Pratt, "UFO Reports Bring a Few Good-Natured Laughs," Abilene *Reporter-News,* January 16, 2008, accessed November 1, 2010, http://www.reporternews.com/news/2008/Jan/16/ufo-reports-bring-a-few-good-natured-laughs/

Bill Radke, "A Sighting in Stephenville," American Public Radio, January 26, 2008, November 1, 2010. http://weekendamerica.publicradio.org/display/web/2008/01/25/ufo/

July Danley, President/CEO Stephenville Chamber of Commerce, Interview, September 2009.

Treva Thompson, then-Marketing and Tourism Dir., Stephenville Chamber of Commerce, Interview, September 2009.

Craig Shelburne, "Jewel, Ty Murray, Live the Small Town Life," *CMT News,* May 30, 2006.

Matt Copeland, Co-owner, Barefoot Athletics, Interview, September 2009.

Katy Copeland, Barefoot Athletics, Interview, September 2009.

Christopher O'Brien, *Secrets of the Mysterious Valley* (Adventures Unlimited Press, 2007): 322–38.

Jean Edwards, Interview, December 2009.

George Edwards, Interview, December 2009.

Bruce Maccabee, Interview, November 2009.

Brian Dunning, "The Rendlesham Forest UFO," *Skeptoid*, January 6, 2009, accessed November 1, 2010, http://skeptoid.com/episodes/4135

Michael Shermer, *How We Believe* (Henry Holt, 2000): 172–73, 202–6.

Philip J. Klass, "Plasma Theory May Explain Many UFOs," *Aviation Week and Space Technology* (August 22, 1966): 48.

Philip J. Klass, "Many UFOs Are Identified as Plasmas," *Aviation Week and Space Technology* (October 3, 1966): 54.

Philip J. Klass, *UFOs Identified* (Random House, 1968).

Martin Shough, "A Social History of Ball Lightning," *Magonia* (May 2003), accessed November 1, 2010, http://magonia.haaan.com/2010/balllightning/

M. Stenhoff, *Ball Lightning: An Unsolved Problem in Atmospheric Physics* (Springer, 1999).

Brian Dunning, "Ball Lightning," *Skeptoid*, February 9, 2010, accessed November 1, 2010, http://skeptoid.com/episodes/4192

Philip J. Klass, "Spaceships or Mirages over Washington National Airport, 1952?" The Klass Files, *The Skeptics UFO Newsletter, Skeptical Inquirer*, July 1, 1998, accessed November 1, 2010, http://www.csicop.org/specialarticles/show/klass_files_volume_52/

James E. McDonald, "Comments of a Researcher: Case 5. Washington National

Airport," July 19 and 26, 1952, accessed November 1, 2010, http://www.nicap.org/wnsmcd.htm

J. McDonald, "UFOs—An International Scientific Problem, Paper Presented at the Canadian Aeronautics and Space Institute Astronautics Symposium, Montreal, Canada," March 12, 1968, accessed November 1, 2010, http://www.ufologie.net/htm/mcdonaldca.htm

Jerome Clark, *The UFO Book: Encyclopedia of the Extraterrestrial* (Visible Ink Press, 2007): 660–62.

J. Moseley, ed., "More About the Late Phil Klass," *Saucer Smear* 52, no. 9, October 20, 2005.

James Randi, *Flim-Flam!* (Prometheus Books, 1982):72–73.

Ronald D. Story, *The Encyclopedia of Extraterrestrial Encounters* (New American Library, 2001): 444–46, 624–26.

Larry King clip, http://www.youtube.com/watch?v=wSkXYmExOnA

Edward de Bono, *I Am Right, You Are Wrong* (Penguin, 1991): 22–30.

I. J. Good, *The Scientist Speculates* (Basic Books, 1963): 15.

David Jones, "Thoughts That Go Pop in the Night," *The Age*, January 13, 1986.

W. Platt, "The Relation of the Scientific 'Hunch' to Research," *Journal of Chemical Education* (October 1931): 1969–2002.

Peter Hessler, *Oracle Bones* (HarperCollins, 2006): 138–47.

Michio Kaku, "Prof Michio Kaku on the Science Behind UFOs and Time Travel," *Telegraph U.K.*, March 20, 2008, accessed November 1, 2010, http://www.telegraph.co.uk/science/science-news/3337049/Prof-Michio-Kaku-on-the-science-behind-UFOs-and-time-travel.html

GERM books, UFO Awareness Day, Philadelphia, PA, July 2009.

Wade Goodwyn, "Air Force Alters Texas UFO Explanation," National Public Radio, January 24, 2008, accessed October 26, 2010. http://www.npr.org/templates/story/story.php?storyId=18375952

Jeffrey Weiss, "Texas UFO Mystery Solved?" *Dallas Morning News,* January 24, 2008, accessed October 26, 2010, http://www.dallasnews.com/sharedcontent/dws/news/localnews/stories/DN-ifos_24met.ART.North.Edition1.3787cf0.html

Angela Brown, "F–16s Were in Area Where UFO Reported," January 23, 2008.

M. Jones, "Report Fuels Spy Plane Theories," BBC News, June 14, 2006, accessed November 1, 2010, http://news.bbc.co.uk/2/hi/programmes/newsnight/5079044.stm

D. Thompson, "What Do the UFO Files Reveal?" *Telegraph U.K.,* August 5, 2010, accessed November 1, 2010, http://www.telegraph.co.uk/news/newstopics/howaboutthat/ufo/7927676/What-do-the-UFO-files-reveal.html

J. Grimston, "Is That a Flying Saucer? No, It's a Stealth Bomber," *Sunday Times*, March 22, 2009, accessed November 1, 2010, http://www.timesonline.co.uk/tol/news/uk/article5950460.ece

Wade Goodwyn, "Dozens Claim They Spotted UFO in Texas," National Public Radio, January 16, 2008, accessed November 1, 2010, http://www.npr.org/player/v2/mediaPlayer.html?action=1&t=1&islist=false&id=18146244&m=18159586

Author's note: See Gellene, "How UFOs" for Kinsel quote and *Times* quotes.

J. Gibson, "Dennis Kucinich's UFO Comments Prove He's Nuts," Fox News, October 31, 2007, accessed November 1, 2010, http://www.foxnews.com/story/0,2933,307117,00.html

M. M. Phillips, "What Kucinich Saw," *Wall Street Journal*, January 2, 2008.

Dave Gilson, "Kucinich's UFO Sighting, What He Really Saw," *Mother Jones*, January 4, 2008, accessed November 1, 2010, http://motherjones.com/mojo/2008/01/kucinichs-ufo-sighting-what-he-really-saw

Merrill Goozner, "Send the Editor into Space," *Huffington Post*, January 2, 2008, accessed November 1, 2010, http://www.huffingtonpost.com/merrill-goozner/answer-send-the-editor-in_b_79224.html

Frank Burke, Interview September 2009.

Phil Patton, "UFO Myths: A Special Investigation into Stephenville and Other Major Sightings," *Popular Mechanics* (March 2009), accessed November 2010, http://www.popularmechanics.com/technology/aviation/ufo/4304170

Glen Schulze, Interview, October 2009.

"The Stephenville Lights: What Actually Happened," *Skeptical Inquirer* 33, no. 1 (January/February 2009), accessed October 26, 2010, http://www.csicop.org/si/show/stephenville_lights_what_actually_happened/

Billy Cox, "Multiple Sources Never Hurt," De Void, blog, *Sarasota Herald-Tribune*, February 11, 2008, accessed October 26, 2010, http://www.heraldtribune.com/article/20080211/BLOG32/76223561

Mark Murphy, Interview, September 2009.

CHAPTER 5: WAS THERE A GHOST IN MY HOUSE?

James Boswell, *The Life of Samuel Johnson* (Penguin Classics, 1979): 239.

Laura Spinney, "We Can Implant Entirely False Memories," *Guardian*, December 4, 2003, accessed November 1, 2010, http://www.guardian.co.uk/science/2003/dec/04/science.research1

S. J. Ceci, "Repeatedly Thinking About a Non-Event," *Consciousness and Cognition* 2 (1994): 388–407.

K. A. Braun, "Make My Memory: How Advertising Can Change Our Memories of the Past," *Psychology and Marketing* 19, no. 1 (January 2002): 23.

E. F. Loftus, "Planting a 30-Year Investigation of the Malleability of Memory," *Learning and Memory* 12 (2005): 361–66.

K. A. Wade, "A Picture Is Worth a Thousand Lies: Using False Photographs to create False Childhood Memories," *Psychonomic Bulletin and Review* 9 (2002): 597–603.

Jean Piaget, *Play, Dreams and Imitation in Childhood* (Norton, 1962): 188.

Author's note: There are a number of resources, including some complicated technical papers, describing typical plumbing problems that can lead people to believe their house is haunted. Here is a short selection of more reader-friendly items and some YouTube videos that capture some common, plumbing-related noises. An aside: I sent these to my father, sister, and oldest brother, who are the living people with the greatest recall of the noises. They all thought the examples here were ridiculously unlike the noise in our house. All the online videos and resources here were accessed on October 26, 2010.

Plumbing Noises: http://www.factsfacts.com/MyHomeRepair/PipeNoises.htm

Waterhammer: A complex phenomenon with a simple solution, http://www.omega.com/techref/waterhammer.html

"Peculiar Pipes: Why That Clanging Noise May Have More to do With Porcelain than Poltergeists," http://eisenmanagementgroup.wordpress.com/2009/05/26/peculiar-pipes-why-that-clanging-noise-may-have-more-to-do-with-porcelain-than-poltergeists/

http://www.youtube.com/watch?v=6sdiEdwUO7A

http://www.youtube.com/watch?v=6HCv1YRB-dk&feature=related

http://www.youtube.com/watch?v=0XchGnLiH_o&feature=related

http://www.youtube.com/watch?v=Snp6CkA0eYA&feature=related

R. T. Carroll, "infrasound," *Skeptic's Dictionary*, accessed October 26, 2010, http://www.skepdic.com/infrasound.html6.

Chris Arnot, "Ghost Buster," *Guardian*, July 11 2000, accessed November 1, 2010, http://www.guardian.co.uk/education/2000/jul/11/highereducation.chrisarnot

V. Tandy and T. Lawrence, "The Ghost in the Machine," *Journal of the Society for Psychical Research* 62, no. 851 (April 1998): 360–64. Accessed November 1, 2010. http://www.psy.herts.ac.uk/ghost/ghost-in-machine.pdf

V. Tandy, "Something in the Cellar," *Journal of the Society for Psychical Research* 64.3, no. 860 (July 2000), accessed November 1, 2010, http://www.psy.herts.ac.uk/ghost/Something-in-the-Cellar.pdf

Mary Roach, *Spook* (Norton, 2005): 227–40.

J. J. Braithwaite and M. Townsend, "Good Vibrations: The Case for a Specific Effect of Infrasound in Instances of Anomalous Experience Has Yet to Be Empirically Demonstrated," *Journal for the Society of Psychical Research* 70, no. 885 (2006): 211–24.

"Infrasound Linked to Spooky Effects," Associated Press, September 7, 2003, accessed October 26, 2010, http://www.msnbc.msn.com/id/3077192/

C. C. French et al., "The 'Haunt' Project: An Attempt to Build a 'Haunted' Room by Manipulating Complex Electromagnetic Fields and Infrasound," *Cortex* 45, no. 5 (2009): 619–29.

Hazel Muir, "Where Do Ghosts Come From?" *New Scientist* (October 2009), accessed October 26, 2010, http://www.newscientist.com/article/mg20427321.200-where-do-ghosts-come-from.html

Sheryl C. Wilson and Theodore X. Barber, "The Fantasy-Prone Personality: Implications for Understanding Imagery, Hypnosis, and Parapsychological Phenomena," *Imagery, Current Theory, Research and Application* (1983): 340–90.

Steven Novella, "The Fantasy Prone Personality," *Neurologica Blog*, April 3, 2007, accessed October 26, 2010 http://theness.com/neurologicablog/?p=84

Joe Nickell, "A Study of Fantasy Proneness in the Thirteen Cases of Alleged Encounters in John Mack's Abduction," *Skeptical Inquirer* 20 no. 3 (May/June 1996), accessed November 1, 2010, http://www.csicop.org/si/show/study_of_fantasy_proneness_in_the_thirteen_cases_of_alleged_encounters_in_j/

Susan Clancy, *Abducted* (Harvard Univ. Press, 2005): 132–53.

Peter Hough, "Alien Abductions Revisited: Study Suggests Alien Abduction Experiences Not Simply Products of Fantasy-Proneness," *Fortean Times* (February 2010), accessed November 1, 2010, http://www.forteantimes.com/features/fbi/2929/alien_abductions_revisited.html

P. Hough and P. Rogers, "Individuals Who Report Being Abducted by Aliens: Investigating the Differences in Fantasy-Proneness, Emotional Intelligence and the Big Five Personality Factors," *Imagination, Cognition and Personality* 27, no. 2 (2007): 139–61.

M. Rodeghier et al., "Psychosocial Characteristics of Abductees," *Journal of UFO Studies* 3 (1991): 59–90.

N. P. Spanos et al., "Close Encounters: An Examination of UFO Experiences," *Journal of Abnormal Psychology* 102, no. 4 (1993): 624–32.

M. Eckblad, "Magical Ideation as an Indicator of Schizotypy," *Journal of Consulting and Clinical Psychology* 51, no. 2 (1983): 215–25.

Joel Fischer, *Measures for Clinical Practice: Adults* (Free Press, 1987): 335–36.

S. J. Lynn and J. W. Rhue, "The Fantasy-Prone Person: Hypnosis, Imagination, and Creativity," *Journal of Personality and Social Psychology* 51, no. 2 (August 1986): 404–8.

Chris French, Interview, February 2010.

B. Colvin, "The Acoustic Properties of Unexplained Rapping Sounds," *Journal of the Society for Psychical Research* 73.2, no. 899 (2010): 65–93.

Society for Psychical Research's announcement, accessed October 26, 2010, http://www.spr.ac.uk/main/news/colvin-acoustic-properties-poltergeist-rapping

Steve Volk, "The Lou Gentile Experience," *Philadelphia Weekly*, September 27, 2006, November 1, 2010, http://www.philadelphiaweekly.com/news-and-opinion/cover-story/the_lou_gentile_experience-38419469.html

CHAPTER 6: TO INFINITY AND BEYOND

K. Lorenz, *On Aggression* (Harvest, 1963): 273.

G. Easterbrook, "Why We Shouldn't Go to Mars," *Time*, January 26, 2004, accessed November 1, 2010, http://www.time.com/time/magazine/article/0,9171,993172,00.html

Edgar Mitchell, Interview, February 2009.

Edgar Mitchell, *The Way of the Explorer* (New Page Books, 2008): 15–16 (short description of epiphany), 24–25 (childhood material, early education), 30 (naval career), 42 (space flight as evolutionary step), 73–77 (longer description of space flight and his epiphany), 102–105 (Norbu Chen story), 197–204 (dyadic model).

Author's note: The material cited here is woven throughout the chapter. For anyone interested in reading more about Mitchell's conception of "dyadic model of the universe, see Mitchell, "Dyadic Model."

W. David Woods, *How Apollo Flew to the Moon* (Praxis, 2008): 176.

B. W. Sibrel, "Astronauts Gone Wild," DVD (AFTH, 2004), accessed November 1, 2010, http://www.youtube.com/watch?v=zZYpfKf3tCc

"Edgar Mitchell's Strange Voyage," *People* 1, no. 6 (1974), accessed October 26, 2010, http://www.people.com/people/archive/article/0,,20063934,00.html

M. Nizza, "When an Astronaut Believes in Aliens," *The Lede* blog, *New York Times*, July 24, 2008, accessed November 1, 2010, http://thelede.blogs.nytimes.com/2008/07/24/when-an-astronaut-believes-in-aliens/

Brian Dunning, "The Astronauts and the Aliens," *Skeptoid,* August 10, 2010, accessed November 1, 2010, http://skeptoid.com/episodes/4218

D. Morrison, "UFOs and Aliens in Space," *Skeptical Inquirer* 33, no. 1 (January/February 2009), accessed November 1, 2010, http://www.csicop.org/si/show/ufos_and_aliens_in_space

Dan Brown, *The Lost Symbol* (Anchor, 2009): 208.

Institute of Noetic Sciences, "IONS Overview," accessed November 1, 2010. http://www.noetic.org/about/overview/

"Secrets of the Lost Symbol," *Dateline NBC,* October 15, 2009.

"Hunting the Lost Symbol," Discovery channel, October 25, 2009.

Barbara Bradley Hagerty, "Woman Reads Dan Brown Novel, Discovers Herself," *All Things Considered*. October 12, 2009, accessed November 1, 2010, http://www.npr.org/templates/story/story.php?storyId=113676181

Rose Welch, executive assistant to the president/board, IONS, Interview, October 2009. (Also Radin, Schlitz.)

Frank White, *The Overview Effect* (American Institute of Aeronautics and Astronautics, 1998): 11–12 (Schweickart's description of space flight), 35–37 ("sensing element" material), 183–85 (interview with Cernan), 187 (Michael Collins quote), 190–91 (Schweickart on the transformative aspects of space flight), 215 (quote from Allen), 247–48 (Garn's experience).

Michael Collins, *Carrying the Fire* (Farrar, Straus and Giroux, 1974): 470.

Association of Space Explorers, accessed November 1, 2010, http://www.space-explorers.org/membership.html

Frank White, Interview, February 2010.

Author's note: Andrew Newberg has agreed to conduct neurological studies of the overview effect.

Uri Geller, Interview, August 2009.

R. Targ and H. Puthoff, "Information Transmission Under Conditions of Sensory Shielding," *Nature* 251, no. 18 (October 1974): 602–7.

A. Harrington, ed., *The Placebo Effect* (Harvard Univ. Press, 1999): 40–54 (good overview of placebo effect), 66 ("voodoo" or negative placebo effect in asthma), 211–16. *Author's note:* Great discussion among experts, with discussion of placebo effect's limits.

R. Kradin, *The Placebo Response and the Power of Unconscious Healing* (Routledge, 2008): 169–97.

Andrew Newberg and Mark Robert Waldman, *Born to Believe* (Free Press, 2006): 3–4.

B. Klopfer, "Psychological Variables in Human Cancer," *Journal of Projective Techniques* (1957): 21.

B. O'Regan and C. Hirshberg, *Spontaneous Remission of Cancer: Epidemiological and Psychosocial Aspects* (Institute of Noetic Sciences, 1993).

Dean Radin, *The Conscious Universe* (HarperOne, 1997): 117–33.

Dean Radin, *Entangled Minds* (Paraview Pocket Books, 2006): 161–80.

Jessica Utts, Interview, August 2009.

Edgar Mitchell, "A Dyadic Model of Consciousness," *World Futures* 46, no. 2 (1996): 69–78.

The Overview Institute, accessed October 26, 2010, http://www.overviewinstitute.org/

Virgin Galactic promotional information, accessed October 26, 2010, http://www.virgingalactic.com/overview/experience/

C. B. Thomas, "The Space Cowboys," *Time*, February 22, 2007, accessed November 1, 2010, http://www.time.com/time/magazine/article/0,9171,1592834,00.html

D. Freedman, "The Future of NASA," *Discover* (September 2006), accessed November 1, 2010, http://discovermagazine.com/2006/sep/cover

Elizabeth Landau, "When Being Turned On Is a Turnoff," CNN, April 17, 2010, accessed November 1, 2010, http://articles.cnn.com/2010–04–17/health/

sexual.arousal.disorder_1_arousal-disorder-orgasm-selective-serotonin-reuptake-inhibitors?_s=P.M.:HEALTH

"Diseases from Space and the Giggle Factor," *Biot Report*, 456, August 25, 2007.

Suburban Emergency Management Project, accessed October 26, 2010, http://www.semp.us/publications/biot_reader.php?BiotID=456

E. J. Lyman, "'Giggle Factor' Is No Laughing Matter to Scientists," *USA Today*, March 11, 2003.

J. Johnson Jr., "His Inn Will Be Way Out," *Los Angeles Times*, August 30, 2006, accessed November 1, 2010, http://articles.latimes.com/2006/aug/30/science/sci-bigelow30

Eliza Strickland, "Scratch a Space Nut, Find a Starry-Eyed Hippie," *Wired*, July 20, 2007, accessed November 1, 2010, http://www.wired.com/science/space/news/2007/07/overview

L. David, "Space Colonization Efforts Quietly Pick Up Steam," *USA Today*, February 23, 2005.

Marilyn Schlitz, Interview, November 2009.

CHAPTER 7: THE OPEN MIND

Barack Obama, Tucson Memorial, January 12, 2011, accessed January 13, 2011, http://www.whitehouse.gov/the-press-office/2011/01/12/remarks-president-barack-obama-memorial-service-victims-shooting-tucson

Joe Henry, "Flag," *Tiny Voices* (Anti, 2003).

Andrew Newberg, Interviews, class lectures conducted through the fall of 2009.

Andrew Newberg has since joined the Myrna Brind Center of Integrative Medicine at Thomas Jefferson University and Hospital, where he can devote more time to research and writing in the area of neurotheology and medicine. He remains affiliated with Penn as an adjunct professor and still teaches a class on Neurotheology.

Andrew Newberg, *Principles of Neurotheology* (Ashgate, 2010).

Andrew Newberg, Michael Persinger, and Vilayanur Ramachandran, *God on the Brain: A Neurological Basis for the Religious Impulse*, DVD featuring Richard Dawkins, (BBC, 2003).

Andrew Newberg et al., *Why God Won't Go Away* (Ballantine, 2001): 1–10.

Andrew Newberg et al., "The Measurement of Regional Cerebral Blood Flow During the Complex Cognitive Task of Meditation: A Preliminary SPECT study." *Psychiatry Research: Neuroimaging* 106 (2001): 113–22.

Andrew Newberg et al., "The Neurophysiological Correlates of Meditation: Implications for Neuroimaging," *Journal of the Indian Academy of Clinical Medicine* 3 (1998): 13–18.

"Robertson's True Story," January 13, 2010, *The 700 Club* (via Media Matters for America), accessed October 26, 2010, http://mediamatters.org/mmtv/201001130024

Laurie Goodstein, "Vatican Declined to Defrock U.S. Priest Who Abused Boys," *New York Times*, March 24, 2010, accessed November 1, 2010, http://www.nytimes.com/2010/03/25/world/europe/25vatican.html

N. Kulish and K. Bennhold, "Memo to Pope Described Transfer of Pedophile Priest," *New York Times*, March 25, 2010, accessed November 1, 2010, http://www.nytimes.com/2010/03/26/world/europe/26church.html

Christopher Hitchens, *God Is Not Great* (Twelve, 2007): 53, 56.

Daniel Dennett, *Darwin's Dangerous Idea* (Simon and Schuster, 1995): 18.

Richard Dawkins, "Has the World Changed?" *Guardian,* October 11, 2001, accessed October 26, 2010, http://www.guardian.co.uk/world/2001/oct/11/afghanistan.terrorism2

Andrew Newberg, *How God Changes Your Brain* (Ballantine, 2009): 24–31 (meditation study, memory, plasticity), 137–41 (prejudice, us and them thinking, fundamentalism), also 131–46 (clash of worldviews), 244–48 (Newberg's own feelings on God).

Andrew Newberg et al., "The Measurement of Regional Cerebral Blood Flow During Glossolalia: A Preliminary SPECT Study," *Psychiatry Research: Neuroimaging* (2006): 67–71.

Gospel According to Al Green, R. Mugge, dir., DVD (Acorn Media, 2009).

J. T. Richardson, "Psychological Interpretations of Glossolalia: A Reexamination of Research," *Journal for the Scientific Study of Religion* 12, no. 2 (1973): 199–207. *Author's note:* This study argues that "maladjustment" should remain a consideration in further studies of glossolalia.

N. Spanos and E .C. Hewitt, "Glossolalia: A Test of the 'Trance' and Psychopathology Hypotheses," *Journal of Abnormal Psychology* 88, no. 4 (August 1979): 427–34.

L. J. Francis et al., "Personality and Glossolalia: A Study Among Male Evangelical Clergy," *Pastoral Psychology*, 51, no. 5 (2003): 391–96.

Religulous, DVD (Lion's Gate, 2009).

S. Harris et al., "The Neural Correlates of Religious and Nonreligious Belief," *Plos One* (2009), accessed November 1, 2010, http://www.plosone.org/article/info%3Adoi%2F10.1371%2Fjournal.pone.0007272 *Author's note*: Though some, including me, occasionally accuse Harris of being too strident by a factor of 72, I find this paper remarkably even-handed in assessing the way our most cherished religious and nonreligious beliefs become a part of us.

Dinesh D'Souza, "Atheism, Not Religion, Is the Real Force Behind the Mass Murders of History," *Christian Science Monitor*, November 21, 2006, accessed October 26, 2010, http://www.csmonitor.com/2006/1121/p09s01-coop.html

Robert A. Pape, *Dying to Win* (Random House, 2005): 208–16.

The Reality Club. *Author's note:* Atran's critique of his fellow atheists can be read, in full, at the Reality Club web site, accessed October 26, 2010,http://www.edge.org/discourse/bb.html

A. Newberg and M. Waldman, *Born to Believe* (Free Press, 2006): 45–65 (human perception), 75–76 (belief formation), 90–95 (limits and benefits of reductionistic thinking).

Author's note: These papers, about the philosophical questions raised by the nature of human perception, are well worth a look. While from an operational perspective, it simply doesn't *work* to go around behaving like our every perception might be in error, the lessons here are worth keeping in mind when we are approaching each other or trying to make our minds up about the nature of reality and human experience.

B. Bennett, "Perception and Evolution," *Perception and the Physical World: Psychological and Philosophical Issues in Perception*, DOI: 10.1002/0470013427, 2002: 229–45.

T. Davies et al., "Visual Worlds: Construction or Reconstruction?" *Journal of Consciousness Studies* 9 (2002): 72–87.

T. Crane, "The Problem of Perception," *Stanford Encyclopedia of Philosophy* (2005), http://plato.stanford.edu/entries/perception-problem/

Author's note: These observations about brain function, belief, and how we evaluate information are supported by my interviews with Newberg and his lectures. Please also see Notes and Sources for chapter 1 for a full selection of papers demonstrating the kinds of automatic brain processes (by no means restricted to the functioning of the amygdala) that influence our cognition and predispose us to emerge from every conversation with our belief system intact. The following papers help prove the point.

J. G. Gunnell, "Are We Losing Our Minds? Cognitive Science and the Study of Politics," *Political Theory* 35, no. 6 (December 2007): 704–31.

E. A. Phelps, "Emotion and Cognition: Insights from Studies of the Human Amygdala," *Annual Review of Psychology* 57 (2006): 27–53.

T. Landis, "Emotional Words: What's So Different from Just Words?" *Cortex* 42, no. 6 (2006): 823–30.

S. Hamman, "Positive and Negative Emotional Verbal Stimuli Elicit Activity in the Left Amygdala," *Neuroreport* 13, no. 1 (2002): 15–19.

T. Frodl et al., "Larger Amygdala Volumes in First Depressive Episode as Compared to Recurrent Major Depression and Healthy Control Subjects," *Biological Psychiatry* 53, no. 4 (February 2003): 338–44.

Y. Matsuoka, "A Volumetric Study of Amygdala in Cancer Survivors with Intrusive Recollections," *Biological Psychiatry* 54, no. 7 (October 2003): 736–43.

E. Yoshikawa, "Prefrontal Cortex and Amygdala Volume in First Minor or Major Depressive Episode After Cancer Diagnosis," *Biological Psychiatry* 59, 8 (April 2006): 707–12.

Ara Norenzayan et al., "Mortality Salience and Religion: Divergent Effects on the Defense of Cultural Worldviews for the Religious and the Non-Religious," *European Journal of Social Psychology* 39 (2009): 101–13.

Author's note: Here is a further selection of studies that illustrate the way human beings, regardless of belief systems, seek to ally themselves with preferred groups and reject members of *out* groups without analysis.

A. Olsson et al., "The Role of Social Groups in the Persistence of Learned Fear," *Science* 309, no. 5735 (July 2005): 785–87.

N. T. Feather, "Acceptance and Rejection of Arguments in Relation to Attitude Strength, Critical Ability and Intolerance of Inconsistency," *Journal Abnormal and Social Psychology* 69, (1964): 127–36.

A. Miller, ed., *The Social Psychology of Good and Evil* (Guilford Press, 2004).

C. A. Insio et al., "Conformity and Group Size: The Concern with Being Right and the Concern with Being Liked," *Personality and Social Psychology Bulletin* 11, no. 1 (March 1985): 41–50.

Mark Robert Waldman, Interviews, January, February, and March, 2009. Like Newberg, Waldman was generous with his time.

Waldman is an Associate Fellow at Newberg's Center for Spirituality and the Mind at the University of Pennsylvania, and he is now adjunct faculty in the Executive MBA Program at Loyola Marymount University, where he teaches communication strategies based on their collaborative experiments.

"Father Thomas Keating—Centering Prayer: Its History and Importance," April 13, 2009, from Ken Wilber's *Integral Options Café,* accessed August 22, 2010, http://integral-options.blogspot.com/2009/04/father-thomas-keating-centering-prayer_13.html

J. R. Carey et al., "Neuroplasticity Promoted by Task Complexity," *Exercise and Sport Sciences Reviews* 33, no. 1 (January 2005): 24–31.

Chris McDougall, *Born to Run* (Knopf, 2009): 145–46.

C. Trageser, "Transcendental Steps (or How I Learned to Love Running Without an iPod)," *Runner's World* (May 2010).

T. Wright, "The Spirit of the Running People: Three Cultures You Should Know," *Vagabondish*, accessed October 26, 2010, http://www.vagabondish.com/running-cultures/

Dean Radin, Interview, October 2009.

R. J. Davidson, "Buddha's Brain: Neuroplasticity and Meditation," *IEEE Signal Processing Magazine* 176 (September 2007).

A. Lutz, "Long-Term Meditators Self-Induce High-Amplitude Gamma Synchrony During Mental Practice," *Proceedings of the National Academy of Sciences,* no. 46 (2004): 16369–373.

R. J. Davidson, "Empirical Explorations of Mindfulness: Conceptual and Methodological Conundrums," *Emotion* 10, no. 1 (2010): 8–11.

Author's note: The material describing how science might teach us what kind of God is most beneficial to us, neurologically, is found in chapter 7 of Newberg, *How God Changes Your Brain*. It's also something I discussed with Mark Robert Waldman and Andrew Newberg.

"1000 Rabbis Warn: Open Homosexuality in the Military Is a Disaster and May Cause Further Natural Disasters," *Christian Newswire,* Contact listed as Rabbi Yehuda Levin, Spokesman, Rabbinical Alliance of America.

J. F. Harris, "God Gave U.S. 'What We Deserve,' Falwell Says," *Washington Post*, September 14, 2001.

J. A. Dusek et al., "Genomic Counter-Stress Changes Induced by the Relaxation Response," *Plos One* 3, no. 7 (2008), accessed November 1, 2010, http://www.plosone.org/article/info:doi/10.1371/journal.pone.0002576.

R. Chari et al., "Effect of Active Smoking on the Human Bronchial Epithelium Transcriptome," *BMC Genomics* (2007), accessed November 1, 2010, http://www.biomedcentral.com/1471–2164/8/297

Author's note: Smalley's site (http://www.suesmalley.com/topics/mindfulness/) provides a good introduction to her work. Also see her collection of mindfulness meditation research at: http://www.mindfulexperience.org/publications.php)

S. L. Smalley, *Fully Present: The Science, Art, and Practice of Mindfulness* (Da Capo, Lifelong Books, 2010).

Patricia Fitzgerald, "What Inspired A Scientist to Open a Meditation Center at UCLA?" *Huffington Post*, July 15, 2009, accessed November 1, 2010, http://www.huffingtonpost.com/dr-patricia-fitzgerald/what-inspired-a-scientist_b_228356.html

Author's note: I find Hitchens's comments on Buddhism particularly instructive in illustrating the degree to which the New Atheism can lead to throwing the baby out with the bathwater—dismissing spirituality along with dogmatic religion. Hitchens, in fact, said this of Buddhism in his own *God Is Not Great*: "Those who become bored by conventional 'Bible' religions, and seek 'enlightenment' by way of the dissolution of their own critical faculties into nirvana in any form, had better take a warning. They may think they are leaving the realm of despised materialism, but they are still being asked to put their reason to sleep, and to discard their minds along with their sandals" (p. 204).

David Adam, "Plan for Dalai Lama Lecture Angers Neuroscientists," *Guardian*, July 27, 2005, accessed November 1, 2010, http://www.guardian.co.uk/world/2005/jul/27/research.highereducation

Gardiner Harris, "For N.I.H. Chief, Issues of Identity and Culture," *New York Times*, October 5, 2009, accessed November 1, 2010, http://www.nytimes.com/2009/10/06/health/06nih.html

Andrew Brown, "Sam Harris and Francis Collins," Andrew Brown's blog, *Guardian*, August 2, 2009, accessed November 1, 2010, http://www.guardian.co.uk/commentisfree/andrewbrown/2009/jul/31/religion-atheism-harris-collins-witchcraft

Jon Hamilton, "The Links Between the Dalai Lama and Neuroscience," November 11, 2005, accessed November 1, 2010, http://www.npr.org/templates/story/story.php?storyId=5008565

Marc Kaufman, "Dalai Lama Gives Talk on Science," *Washington Post*, November 13, 2005, accessed November 1, 2010, http://www.washingtonpost.com/wp-dyn/content/article/2005/11/12/AR2005111201080.html

CHAPTER 8: THE IMPOSSIBLE DREAM

T. E. Lawrence, *The Seven Pillars of Wisdom* (Wordsworth Editions, 1997): 7.

Most of the material here was gathered at the Lucidity Institute's March 2010 workshop on the big island of Hilo, Hawaii.

J. Horgan, "*Inception* Is a Clunker, but Lucid Dreaming Is Cool," *Cross Check* blog, *Scientific American* (August 2, 2010), accessed November 1, 2010, http://www.scientificamerican.com/blog/post.cfm?id=inception-is-a-clunker-but-lucid-dr-2010-08-02

Stephanie Rosenbloom, "Living Your Dreams, in a Manner of Speaking," *New York Times*, September 23, 2007, accessed November 1, 2010, http://www.nytimes.com/2007/09/16/fashion/16lucid.html

T. W. Rinpoche, *The Tibetan Yogas of Dream and Sleep* (Snow Lion Publications, 1998).

Aristotle, *On Dreams*, Part III, accessed November 1, 2010, http://classics.mit.edu/Aristotle/dreams.html

H. Saint-Denys, *Dreams and How to Guide Them* (trans., Duckworth, 1982).

C. M. Den Blanken, "An Historical View of Dreams and the Ways to Direct Them . . . ," *Lucidity Letter* 7 (December 1988), accessed November 1, 2010, http://www.spiritwatch.ca/anhis.htm.

F. van Eeden, "A Study of Dreams," *Proceedings of the Society for Psychical Research* 26 (1913), 431–61, accessed October 26, 2010, http://www.lucidity.com/vanEeden.html.

Stephen LaBerge and Howard Rheingold, *Exploring the World of Lucid Dreaming* (Ballantine, 1990): 13–15 (lucid dreaming and waking life), 23–24 (effort to scientifically verify lucid dreaming), 27 (machines malfunction), 39 (unusual things in real life), 40 (reading as a state test, digital watch), 61–64 (testing the dream/waking states).

Jeff Warren, *The Head Trip* (Random House, 2007): 121–23, 213–15 (visualization studies, neuroplasticity).

S. LaBerge et al., "Lucid Dreaming Verified by Volitional Communication," *Perceptual and Motor Skills* 52 (1981): 727–32.

Author's note: LaBerge's story of academic disinterest, or active suppression, was shared in Hawaii. I checked the PubMed database for myself in August 2010 and indeed, page after page of the 1981 issue in which LaBerge published his findings is available—but not his article.

Also, the informed reader should know that in the 1970s British researcher Keith Hearne was concurrently working with lucid dreamer Alan Worsley to establish the validity of the lucid dream.

R. Stephenson et al., "Prolonged Deprivation of Sleep-Like Rest Raises Metabolic Rate in the Pacific Beetle Cockroach," *Journal of Experimental Biology* 210 (2007): 2540–47.

T. Yokogawa et al., "Characterization of Sleep in Zebrafish and Insomnia in Hypocretin Receptor Mutants," *PLoS Biol* 5, no. 10 (October 2007).

T. L. Lee-Chiong, *Sleep* (Wiley-Liss, 2005): 13–16.

J. D. Payne, "Sleep, Dreams, and Memory Consolidation," *Learning and Memory,* 11 (2004) 671–78.

R. D. Cartwright, "The Role of Sleep in Changing Our Minds: A Psychologist's Discussion of Papers on Memory Reactivation and Consolidation in Sleep," *Learning and Memory* 11 (November 2004): 660–63.

F. Crick, "The Function of Dream Sleep," Nature Publishing Group 304, 5922, (1983): 111–14.

Daniel Williams, "While You Were Sleeping," *Time,* April 5, 2007, accessed November 1, 2010, http://www.time.com/time/magazine/article/0,9171,1606872,00.html

Hara Estroff Marano, "Why We Dream," *Psychology Today* (March 1, 2005), accessed October 26, 2010, http://www.psychologytoday.com/articles/200504/why-we-dream

Jim Horne, *Sleepfaring* (Oxford Univ. Press, 2007): 66–67, 182.

A. Revonuso, "The Reinterpretation of Dreams: An Evolutionary Hypothesis of the Function of Dreaming," *Behavioral and Brian Sciences* 23,(2000): 677-901.

O. B. Ramsay and A. Rocke, "Kekulé's Dreams: Separating the Fiction from the Fact," *Chemistry in Britain* 20 (1984): 1093–94.

D. M. Locke, "The Putative Purity of Science," *Science as Writing* (Yale Univ. Press, 1992): 133–66.

Paul Strathern, *Mendeleyev's Dream: The Quest for the Elements* (Thomas Dunne Books, 2000): 285–92.

U. Weiss, R.A. Brown, "An Overlooked Parallel to Kekule's Dream: The Discovery of the Chemical Transmission Of Nerve Impulses by Otto Loewi," *Journal of Chemical Education* 64, no. 9 (1987): 770.

J. van Gijin, "Book Review: The Chemical Languages of the Nervous System: History of Scientists and Substances," *New England Journal of Medicine* 355 (November 2006): 2266–67.

George K. York III, "Otto Loewi: Dream Inspires a Nobel-Winning Experiment on Neurotransmission," *Neurology Today* 4, no. 12 (December 2004): 54–55.

S. Krippner et al., *Extraordinary Dreams and How to Work with Them* (State Univ. of New York Press, 2002): 24.

Mark Blagrove, "Scripts and the Structuralist Analysis of Dreams," *Dreaming: Journal of the Association for the Study of Dreams* 2, no. 1 (March 1992): 23–38.

D. Barrett, "The 'Committee of Sleep,'" *Dreaming: Journal of the Association for the Study of Dreams* 3, no. 2 (1993), accessed November 1, 2010, http://www.asdreams.org/journal/articles/barrett3–2.htm

D. Pick and L. Roper, *Dreams and History* (Routledge, 2004): 159–71.

D. Drabelle, "In Dreams Begin Discoveries," *Pennsylvania Gazette* (January/February 2009), accessed November 1, 2010, http://www.upenn.edu/gazette/0109/feature3_1.html

Author's note: Artists and writers and filmmakers long inspired by their dreams include auteurs like David Lynch and more mainstream, popular figures like Stephanie Meyer, who divined the story for *Twilight* from a dream. Paul McCartney dreamt the music that became "Yesterday," and Keith Richards woke up with "(I Can't Get No) Satisfaction" in his head. Novels like *The Strange Case of Dr. Jekyl and Mr. Hyde* and *Frankenstein* also sprang directly from dreams.

CHAPTER 9: AFTER-DEATH COMMUNICATION?

William James, *The Will to Believe and Other Essays in Popular Philosophy* (Dover, 1956): 29–30.

Tom Lareau, Interview, January 2005.

Allan Botkin, Interviews, February 2005, August 2010.

"About Induced ADCs," from Botkin's own web site, makes the claim of providing IADCs to thousands, http://induced-adc.com/

Allan Botkin, Interview with George Noory, *Coast to Coast A.M.*, October 27, 2004, http://www.coasttocoastam.com/show/2004/10/27. Accessed November 1, 2010.

F. Shapiro, *EMDR* (Basic Books, 1997): 8–10 (discovery), 26–28 (no understood mechanism), 5 (too good to be true), 91–92, 135–36 (information processing).

F. Shapiro, "Eye Movement Desensitization: A New Treatment for Post-Traumatic Stress Disorder," *Journal of Behavior Therapy and Experimental Psychiatry* 20, no. 3 (1989): 211–17.

R. T. Carroll, "Eye Movement Desensitization and Reprocessing (EMDR)," *Skeptic's Dictionary,* accessed October 27, 2010, http://www.skepdic.com/emdr.html

F. Shapiro, "Efficacy of the Eye Movement Desensitization Procedure in the Treatment of Traumatic Memories," *Journal of Traumatic Stress Studies* 2 (1989): 199–223.

A. Ehlers et al., "Do All Psychological Treatments Really Work the Same in Posttraumatic Stress Disorder?" *Clinical Psychology Review* 30, no. 2 (2010): 269–76.

Author's note: Here is a by no means complete listing of studies attesting to EMDR's effectiveness.

B. van der Kolk, "The Psychobiology and Psychopharmacology of PTSD," *Human Psychopharmacology: Clinical and Experimental* 16 (2001): 49–64.

G. H. Seidler, "Comparing the Efficacy of EMDR and Trauma-Focused Cognitive-Behavioral Therapy in the Treatment of PTSD: A Meta-Analytic Study," *Psychological Medicine* 36, no. 11 (2006): 1515–22.

R. Rodenburg, "Efficacy of EMDR in Children: A Meta-Analysis," *Clinical Psychology Review* 29, no. 7 (November 2009): 599–606.

M. L. Van Etten, "Comparative Efficacy of Treatments for Posttraumatic Stress Disorder: A Meta-Analysis," *Clinical Psychology and Psychotherapy* 5 (1998): 126–44.

C. M. Chemtob et al., "Eye Movement Desensitization and Reprocessing," *Effective Treatments for PTSD: Practice Guidelines from the International Society for Traumatic Stress Studies* (Guilford, 2009): 283–301.

"Eye Movement Desensitization and Reprocessing (EMDR) Treatment for Psychologically Traumatized Individuals," *Effective Treatments for PTSD: Practice Guidelines from the International Society for Traumatic Stress Studies* (Guilford, 2000): 333–35.

S. A. Wilson et al., "Eye Movement Desensitization and Reprocessing," *Journal of Consulting and Clinical Psychology* 63 (1995): 928–37.

S. A. Wilson et al., "15-Month Follow-Up of Eye Movement Desensitization and Reprocessing (EMDR) Treatment for Psychological Trauma," *Journal of Consulting and Clinical Psychology* 65, no. 6 (1997): 1047–56.

J. G. Carlson, "Eye Movement Desensitization and Reprocessing (EMDR) Treatment for Combat-Related Posttraumatic Stress Disorder," *Journal of Traumatic Stress* 11, no. 1 (January 2008): 3–24.

Robbie Dunton, EMDR Institute, Interview, November 2010.

A fuller listing of EMDR's laurels can be obtained at the EMDR International Association web site: http://www.emdria.org/

Allan Botkin and R. Craig Hogan, *Induced After-Death Communication* (Hampton Roads, 2005): 10–15.

R. Stickgold, "EMDR: A Putative Neurobiological Mechanism of Action," *Journal of Clinical Psychology* 58, no. 1 (2002): 61–75.

C. T. Smith, "Posttraining Increases in REM Sleep Intensity Implicate REM Sleep in Memory Processing and Provide a Biological Marker of Learning Potential," *Learning and Memory* 6 (2004): 714–19.

Thomas Mellman, "REM Sleep and the Early Development of Posttraumatic Stress Disorder," *American Journal of Psychiatry* 159 (October 2002): 1696–1701.

K. Lansing, et. al, "High-Resolution Brain SPECT Imaging and Eye Movement Desensitization and Reprocessing in Police Officers With PTSD," *Journal of Neuropsychiatry and Clinical Neurosciences,* 17, no. 4, (2005): 526–532.

P. Levin, P., et. al, "What psychological testing and neuroimaging tell us about the treatment of posttraumataic stress disorder by eye movement desensitization and reprocessing," *Journal of Anxiety Disorders,* 13, no. 1–2 (1999): 159–172

B. van der Kolk, "The psychobiology of traumatic memory: Clinical implications of neuroimaging studies," *Annals of the New York Academy of Sciences,* 821 (1997): 99–113.

I interviewed six soldiers in one group and had individual interviews with three more. One soldier declined to have his last name published. Thanks to Jimmy Rivers, Pete Reed, Wendell Marks, Ramone Calderon, George, Mike Sylvia, Mike Dick, Paul Thomas, and Tom Lareau.

Induced After-Death Communications, under "Trained Therapists Available Today" at http://induced-adc.com/

The IADC therapists I interviewed included Katelynn Daniles, Greg Rimoldi, and Hania Stromberg, Ph.D.

Bessel van der Kolk, Interview, March 2005.

As I was finishing this book, Botkin said he has found a researcher who is interested in conducting a study on the effectiveness of IADC therapy.

CHAPTER 11: OUR TIME IN HELL

William James, "The Energies of Men," first published in *Science,* no. 635 (1907): 321–32, accessed October 27, 2010, http://psychclassics.yorku.ca/James/energies.htm

Lucid Dreaming Workshop, March 2010.

Stephen LaBerge, *Exploring the World of Lucid Dreaming* (Ballantine, 2004): 87.

Robert Waggoner, *Lucid Dreaming: Gateway to the Inner Self* (Moment Point Press, 2009): 51–75, 158–59 (Keelin story).

"Conversation Between Stephen LaBerge and Paul Tholey in July of 1989." *Author's note:* This interview took place at the 1989 Association for the Study of Dreams (ASD) conference in London. This interview, available online, nicely illustrates LaBerge's position that the other dream figures are in fact our own mental constructs (or at least there is no proof of their independence). Accessed October 27, 2010, http://www.futurehi.net/docs/Laberge_Tholey.html

Sam Harris, *The End of Faith* (Norton, 2004): 40–41.

INDEX